PROCEEDINGS OF THE FOURTH INTERNATIONAL CONFERENCE
ON THE APPLICATION OF STRESS-WAVE THEORY TO PILES
THE HAGUE/NETHERLANDS/21-24 SEPTEMBER 1992

Application of Stress-Wave Theory to Piles: Test Results

A.A.BALKEMA/ROTTERDAM/BROOKFIELD/1996

The texts of the various papers in this volume were set individually by typists under the supervision of either each of the authors concerned or the editor.

Authorization to photocopy items for internal or personal use, or the internal or personal use of specific clients, is granted by A.A. Balkema, Rotterdam, provided that the base fee of US$1.50 per copy, plus US$0.10 per page is paid directly to Copyright Clearance Center, 222 Rosewood Drive, Danvers, MA 01923, USA. For those organizations that have been granted a photocopy license by CCC, a separate system of payment has been arranged. The fee code for users of the Transactional Reporting Service is: 90 5410 846 0/96 US$1.50 + US$0.10.

Published by
A.A. Balkema, P.O. Box 1675, 3000 BR Rotterdam, Netherlands (Fax: +31.10.413.5947)
A.A. Balkema Publishers, Old Post Road, Brookfield, VT 05036-9704, USA
(Fax: +1.802.276.3837)

ISBN 90 5410 846 0
© 1996 A.A. Balkema, Rotterdam
Printed in the Netherlands

Contents

Foreword

During the 3rd International Stress Wave Conference, in Ottawa Canada, the idea was launched to have the 4th International Stress Wave Conference in the Netherlands. After much international pressure the Dutch decided to accept the challenge. An enthusiastic Executive Organising Committee was formed, with representatives from the Dutch foundation industry.

From the beginning the EOC planned to combine the Conference with a Pile Testing Research Project and a Demonstration Day. The organising committee was very pleased that Prof. Bram van Weele was willing to co-ordinate these activities and a Test Organising Committee (TOC) was formed. The organisation and subsequent interpretation and publishing proved to be a much larger task than the members of the TOC ever realised.

More than 200 participants from all continents came to the Conference and the Demonstration Day. Both events turned out very successful.

This book presents the Pile Testing Research Project. Although it covers most of the results, not all recorded data are presented. Several instrumentation activities of the research project were done by companies at their own costs. The intent to publish results in additional reports and papers.

I would like to thank all participants, members of the committees and sponsors for their efforts in making the Conference and Demonstration Day a success. I also would like to thank Messrs. Wim Heijnen and Kees Joustra for their reviewing and editing activities of the book and Prof. Bram van Weele for his constant enthusiasm in motivating the reporters.

I am sure that the secrets of pile testing have a fascination for most participants of Stress Wave Conferences. I therefore hope that the 4th International Stress Wave Conference, the Pile Testing Research Program and Demonstration Day assisted in removing some of these secrets.

Peter Middendorp
Chairman of the 4th International Stress Wave Conference

CHAPTER 1

Introduction

A.F. VAN WEELE

The organisation committee decided that, apart from the theoretical aspects, also the practical side of pile driving and pile testing should receive proper attention during the Conference. This resulted in the formation of a 'Test Organising Committee', which was requested to look into this. The committee made an inventory of subjects and came up with a detailed proposal on how and where such a Test Day could be organised and which topics needed closer attention. Their proposals were accepted and organisation of the day was commenced. This proved to be a huge task, especially the presentation of the results obtained. The Committee decided that all aspects related to the various topics of the Stress-wave Conference should be included in the program. The Committee also decided that this opportunity of close cooperation between so many specialist should be used to perform important and divers research work.

1.1 PREPARATIONS

The Test Day was to be held at a suitable location with a uniform soil profile. This was found in the city of Delft, within the parking area of the main Auditorium of the Technological University. This location had the advantage that adequate facilities for luncheon were available, as well as meeting rooms, so that under adverse weather conditions the show could go on for the participants. The weather was good however, and the outdoor operations were well attended.

Long before the Conference took place, site investigations were carried out and 10 prefabricated test piles, 9 of which had a well defined anomaly, were carefully installed in a row, so that the each of the pile heads could be made easily accessible, approx. 1 m above site-level. In addition some standard prefabricated piles had been installed some weeks before the conference, ready to be statically tested in compression. This information was required during the Conference in order to be able to judge the predictions of the ultimate capacity of similar piles in the same soil profile, based on PDA-analysis, Statnamic testing and Pseudo-static testing. All three forms of testing were carried out during the Test Day as a demonstration for the participants. In Figure 1.1 the siteplan is presented in Figure 1.2 the time schedule.

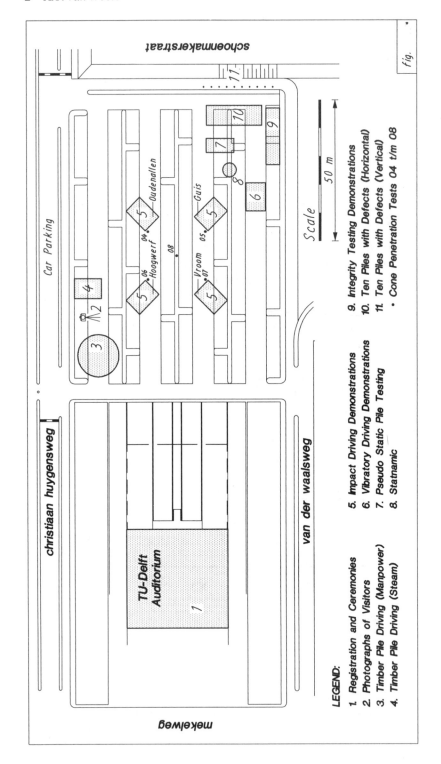

Figure 1.1.

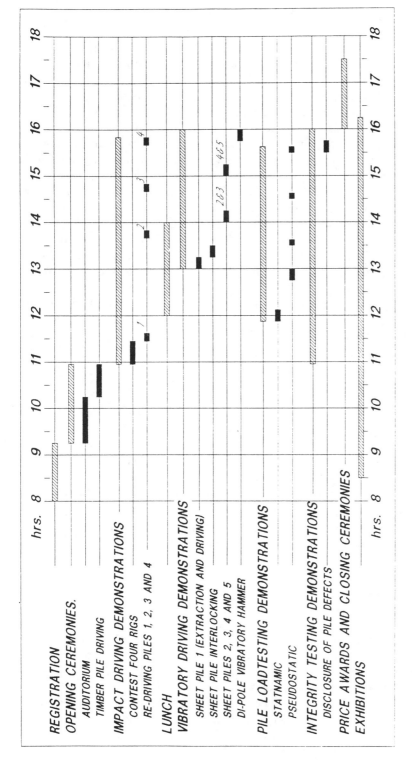

Figure 1.2.

1.2 DRIVING DEMONSTRATIONS.

The driving competition was very important, held between 4 pile-drivers, each equipped with a different hammer and driving closely together 4 identical, slender prestressed concrete piles (20 m long and 250×250 mm cross-section). The competition aimed at assessing speed of driving, noise- and vibration emission, energy consumption and stress generation in the piles. The official start of the driving was during the morning of the Demonstration Day, performed by the Dean of the University with a starting pistol, made it very clear to all participants that demonstrations had begun. From then onwards all kinds of activity took place throughout the area. An antique hand operated pile driver was available to drive timber piles. Most of the representations from the different countries demonstrated their practical ability in manual pile-driving by close cooperation.

The results of the stress-measurements showed after evaluation that the maximum compressive stresses in a pile during driving have no relation whatsoever to the soil resistance, but are governed only by the peak velocity of the hammer upon impact. Conversely the pile reaching the preset foundation level first, arrived there with the lowest compressive stresses of the four!

During the final 1 m of driving, the noise level of each of the 4 pile drivers was monitored independently and vibrations of the ground surface were measured at increasing distances from the pile. During driving of one of the test-piles the pore water pressures in the bearing sand was monitored at a very short distance from the piletoe. The test was set up in such a way that many readings could be taken per blow, so that the dynamic behaviour was recorded.

1.3 LOAD TESTING

A Statnamic Test was also demonstrated, while the Pseudo-static Pile Tester did a series of tests until reaching the ultimate capacity of a pile, every full hour. Both tests take little time, are much less expensive than static testing and are able to supply valuable information about the ultimate capacity.

Most of the results, obtained during the Test Day have been included in this book. Special attention has been paid to te next mentioned items.

1.4 INTEGRITY TESTING

Apart from demonstrations during the Test Day, a competition was held prior to the start of the Conference, with the results of the competition and the correct answers made public during the day. Identical piles, as had been included in the test, were placed above ground in horizontal position, hidden below tarpaulins. These piles could be tested by any of the participants to see if they could recognize which pile had which anomaly. The outcome was presented on a bill-board and at the end of the day the correct answers were revealed and the winner of the day made known. This event attracted considerable attention from colleagues from all over the worl both during and after the Conference. Several of them had difficulty with the fact, that piles with rather serious defects had been installed, which could not be uniquely identified. It was their view that this should not have been done during such a competition. The organisers were however of the opinion that this approach would make impression, to the extent that the market would become better aware of the limitations of the system.

1.5 PREDICTION OF DRIVING BEHAVIOUR

Predicting the driving behaviour of precast concrete piles. The details of the pile to be driven and of the 4 different hammers used, together with details about the driving caps (with dolley and soft wood packings) were given well before the conference took place, upon the request of the participants, in order to enable them to take part in the driving competition. As the driving behaviour results became available immediately, the winner could be announced at the end of the Test Day. Happily enough, all piles reached their intended foundation level, although the piles were deliberately fabricated to be very slender and had to undergo rather hard and prolonged driving. Integrity testing afterwards did not detect damage to any of the piles.

1.6 DRIVING AND EXTRACTING SHEET PILES

A short wall consisting of a numer of interlocked steel sheet piles had been driven some weeks prior to the start of the conference. These sheetpiles were extracted and redriven a short distance away during the testday, using the latest design vibratory drivers, operating at high- and low frequencies. Special attention was paid to the use of 'joint-detectors'. Such detectors are attached to the forward joint, near to the foot of a sheet pile. It reacts when the base of the following section passes and switches on a light near the driver. This demonstration met difficulties and it was only close to the end of the day before the system functioned properly. Vibrations were monitored during driving as well as during extraction by both types of vibrator. One of them was an 'resonance free-vibrator'. This means that a static eccentric moment is gradually applied and removed while the vibrator is running at full speed. This has as important advantage that the peak velocities during switching-off and on can be entirely avoided. Practice has shown that such a vibrator is able to reduce peak velocities by almost 50%.

At the same the Giken-made silent piler was demonstrated. With this tool, driving and extraction of single sheets is made possible, while avoiding vibrations entirely combined with low noise-emission.

1.7 CONE PENETRATION TESTING

Different types of mobile cone-penetrometer units showed how far CPT-technology has developed in The Netherlands. Data acquisition is predominantly fully computer-controlled with readings transmitted digitally for automatic and immediate processing in the office. Many consultants in The Netherlands determine pile depth versus pile capacity at each CPT-location by computer in such a way, that is in compliance with current Dutch Code of Practice. This highly developed practice contributes largely to economic, fast and reliable pile foundation design, unmatched anywhere else in the world.

1.8 COMPANY PRESENTATIONS

Many specialist foundation companies, active in the field of foundation design or foundation construction, demonstrated their abilities during the Conference. Although the programm had much to offer and, during every hour a special event was held, time was adequate for the participants to meet and get acquainted with many Dutch-, and even a few foreign specialist companies.

Looking back at the Demonstration Day the organizers believe that it was a great success and the participants were happy with the many presentations. It contributed to the main goal of conferences such as this, such that participants from all over the world can familiarize themselves with the latest State of the Art techniques and to meet colleagues, active in the same field in order to exchange experience and ideas. The initiative should get a follow up during next conferences as our professional attention is often too much focussed on theoretical aspects and unsufficiently attracted by what is going on in the field. The organizers of the Testday regret very much, that the publication of this book has taken so much more time than originally foreseen. They are very grateful to the many private Dutch companies for their cooperation and contribution, without which the demonstration day could not have been held and this book would not have been published. A list of the participating companies has been added. The Dutch Ministery of Economic Affairs need our sincere thanks for their financial support of the research activities, performed prior to and during the conference.

LIST OF MAIN SPONSORS OF THE TEST DAY

Ballast Nedam B.V. P.O. Box 391 7300 AJ Apeldoorn	Organisation
Guis Funderingstechnieken B.V. D. Duyvisweg 48 3316 BL Dordrecht	Driving Test Pile
P. & G. Hooghwerff B.V. P.O. Box 5723 3290 AA Strijen	Driving Test Pile
International Construction Equipment ICE Hefbrugweg 6 1332 AN Almere	Organisation and demonstration vibratory pile driver

IFCO Foundation Engineering Expertise P.O. Box 334 2800 AH Gouda	Organisation and various monitoring techniques
IHC-Hydroblok B.V. P.O. Box 26 2960 AA Kinderdijk	Organisation
Franki Grondtechnieken B.V. P.O. Box 55 4900 AB Oosterhout	Load Testing Precast Pile
Nederhorst Grondtechnieken B.V.-NGT P.O. Box 303 2800 AH Gouda	Organisation and Demonstration Lock Integrity Indicator
Oudenallen Heibedrijf B.V. P.O. Box 186 3440 AD Woerden	Driving Test Pile
Prepal=Joint Precast Pile Manufactorers P.O. Box 194 3440 AD Woerden	Supply precast concrete piles
Royal Institution of Engineers (KIVI) P.O. Box 30424 2500 GK 's-Gravenhage	Organisation and donation
Van Splunder B.V. Prinsenhof 36 3481 HB Harmelen	Demonstration of Lock Proximity Indicator
Terracon Funderingstecniek B.V. Rijksweg 242 4255 GS Nieuwendijk	Installation IT-test piles
TNO-Bouw P.O. Box 49 2600 AA Delft	Organisation, demonstration Statnamic and various monitoring techniques
Funderingstechnieken Verstraeten B.V. P.O. Box 55 4500 AB Oostburg	Static Load Testing pile and Pseudo Static Pile Load tester demonstrations
Vroom-HBF P.O. Box 7 1474 ZG Oosthuizen	Driving Test Pile
Woud Wormer B.V. Veerdijk 76 1531 MA Wormer	Demonstration vibratory driving and extration

Figure 1.3. General assembly in main auditorium during the opening session. In the front row from left, P. Middendorp, Chairman Organising Committee, Prof. Drs. P.A. Schenk, Dean of the University, Prof. A.F. van Weele, Chairman Test Day Committee and Prof. Dr. F.B.J. Barendsen, Chairman Scientific Committee.

Figure 1.4. Original old Dutch piling rig; free fall hammer operated by Civil Engineering students, assisted by the Dean of the University.

Figure 1.5. Piling forman going to detangle the main pulling rope, which got stuck aside of the top sheave.

Figure 1.6. Participants from South East Asia are driving a timber pile, the old fashioned way.

Figure 1.7. Driving with a drop hammer, operated by an original, steam driven winch. From right to left, piling operator, Dean of the University and Prof. van Weele.

Figure 1.8. Four rigs ready to begin the driving-competition upon a token by the Dean. Far right the rig, equipped with a piling vibrator to drive and extract sheetpiles. In between the PSPL-tester.

Figure 1.9. Oudenallen's piling rig equipped with the Delmag diesel hammer.

Figure 1.10. Ballast-Guis' piling rig with Menck hydraulically operated hammer.

Figure 1.11. Piling rig of HBF-Vroom with IHC-hydrohammer.

Figure 1.12. Piling rig of Hooghwerff equipped with a double acting ICE-diesel hammer.

Figure 1.13. The end of the driving contest is nearby.

Figure 1.14. The test pile driven with the Menck hammer just reached its final depth. Tension among those involved, among them Mr. P. Middendorp of TNO.

Figure 1.15. One of the automatic vibration monitoring systems of IFCO, recording ground vibrations during driving.

Figure 1.16. Elaborate instrumentation of the head of one of the 4 prestressed concrete piles.

Figure 1.17. The silent piler of Van Splunder and the vibratory driver of Woud Wormer, both ready to install steel sheetpiles.

Figure 1.18. The Statnamic Test from TNO/Birminghammer, is getting attention from participants.

Figure 1.19. The Pseudo Static Load Testing apparatus from Verstraeten performed a full scale load test every hour.

Figure 1.20. Foto-exhibition of preparatory works draws attention in the Auditorium.

Figure 1.21. The exposition gets also attention from the participants, among them Prof. Dr. A. Verruit at right.

Figure 1.22. Iintegrity testing of precast piles with a built-in defect for those participants, who did like to have a try. Near the end of the Test Day the piles were exposed and compared with the participant's predictions.

CHAPTER 2

General soil conditions of the test site

W.J. HEIJNEN
Delft Geotechnics, Delft, Netherlands

ABSTRACT: In order to supply soils data to the participants in the different competitions regarding the driving and the prediction of pile behaviour on the basis of dynamic and kinetic pile load tests, soil investigations were carried out at the test site just behind the Auditorium of Delft Technological University. In the following a comprehensive description of the various tests is given. The results are discussed briefly.

2.1 SOIL INVESTIGATIONS

The plan view of the test site is shown in Figure 2.1. The locations of the various tests are indicated.

Ten CPT's and one continuous, diameter 66 mm, type Begemann boring, were carried out prior to the pile driving and testing. The results of these test have been put at the disposal of the participants. All CPT's have been performed in accordance with the standard procedure as described in the report of Technical Committee 16 of the International Society of Soil Mechanics and Foundation Engineering (Swedish Geotechnical Society, 1989).

At the location of each of the testpiles 1 through 5, which were intended for the comparison of the predicted behaviour on the basis of the results of dynamic and kinetic load tests with the results of static load tests, a CPT was made before the installation of these piles.

Figure 2.2 shows the results of the continuous boring and the CPT 08 which was performed in the vicinity of the boring.They were situated at the centre of the test site.

In the diagrams of Figure 2.3 the results of CPT 01C2, made at the south east corner of the test site, are presented.

After the execution of the dynamic and kinetic tests, four additional CPT's were made in the vicinity of the test piles, which were used for the prediction of their behaviour on the basis of the dynamic and kinetic load tests, in order to investigate whether the sandy soil layers had gained strength as a result of the driving the piles or not.

Figure 2.4 shows the result of one of these additional CPT's.

19

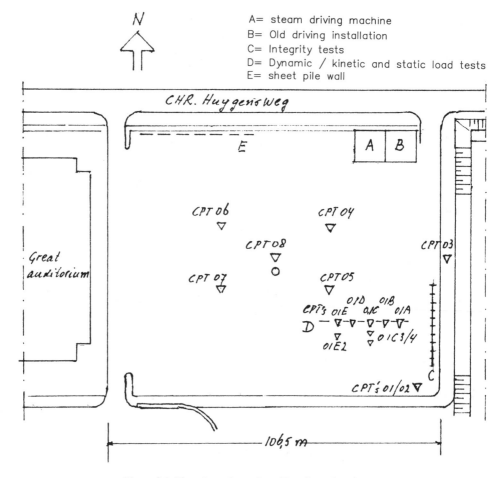

Figure 2.1. Plan view of test site with soil test locations.

2.2 DISCUSSION OF THE RESULTS OF THE SOIL INVESTIGATIONS

The CPT's shown in the Figures 2.2 and 2.4 are to a certain extent, representative for the soil conditions at the test site. The variation across the site in depth of the various layers as well as the strength, are rather insignificant. In particular the cone resistance pattern of the pleistocene layers from 15 to 30 m shows only minor variations. However CPT 01C2, shown in Figure 2.3, which was made in the south east corner of the test site, shows lower q_c-values in comparison with the other CPT's. The difference amounts to approximately 25% of the average q_c-values in the pleistocene sand layer above the depth of 30 m below groundlevel.

  The following approximate soil profile results from the analysis of the continuous boring and the CPT's:

a) Groundlevel to approx. −1 m: fill, sandy and clayey;

b) −1 m to −2,20 m: clay, q_c-value approx. 0,5 to 0,8 MPa;

c) −2,20 m to −5,00 m: silty sands, maximum q_c-values 8 to 10 MPa;

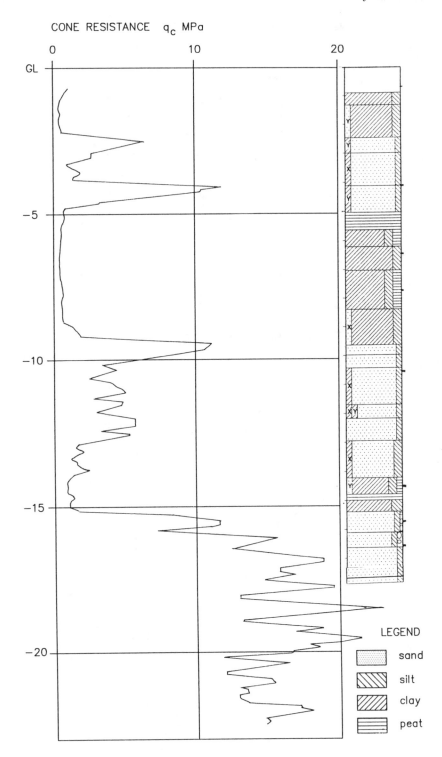

Figure 2.2. Continuous boring and CPT 08.

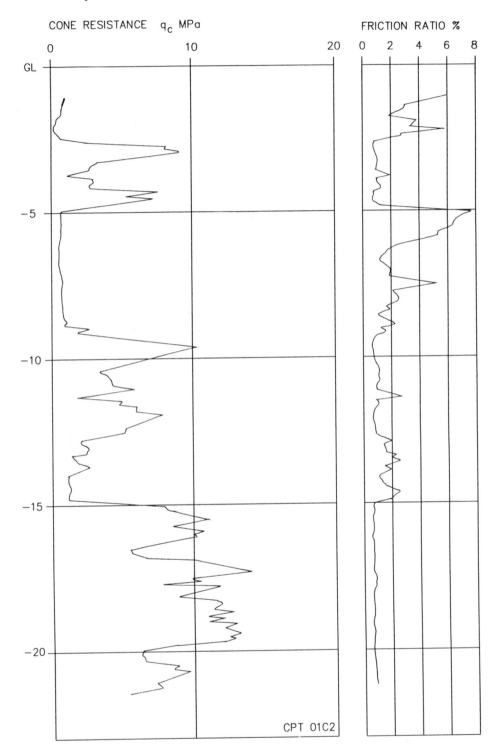

Figure 2.3.

CONE RESISTANCE q$_c$ MPa

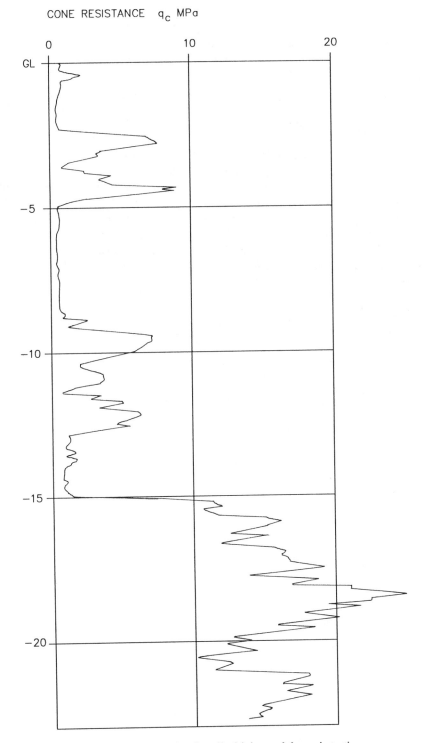

Figure 2.4. CPT 01 E2 - made after pile driving and dynamic testing.

d) –5,00 m to –9,00 m: clay and peat, q_c-value 0,3 to 0,5 MPa;

e) –9,00 m to –14,00 m: silty sand, maximum q_c-values ranging from 6 to 9 MPa;

f) –14,00 m to –15,00 m: clay and peat, q_c-values 1 to 2 MPa;

g) –15,00 m to –>30 m: pleistocene sand, maximum q_c-value at depth of 19 to 20 m below ground level, 22 MPa.

2.3 CONCLUSIVE REMARKS

The soil condition found at the test site is normal for this part of the city of Delft. All buildings in this area are supported by pile foundations.

With the exception of the south east corner at the location of CPT 01C2, soil conditions of the site were considered to be quite uniform and suitable for the anticipated driving tests and the dynamic and kinetic load test competitions.

It is very unlikely that the 'bad' CPT 01C2 had an adverse influence on the results of the comparison between the predictions made on the basis of the dynamic and kinetic tests and the results of the static loading tests because all CPT's at the location of the test piles 1 to 5 were uniform.

REFERENCES

Swedish Geotechnical Society and the Swedish Geotechnical Institut, 1989. Report of the ISSMFE Technical Committee on Penetration Testing of Soils – TC 16, 'Reference Test Procedures CPT - SPT - DP - WST'.

CHAPTER 3

Pile integrity tests

M.Th.J.H. SMITS
Fugro B.V., Leidschendam, Netherlands

3.1 INTRODUCTION

Pile integrity tests were carried out by twelve companies on ten precast concrete piles at the test site at Delft University of Technology. The results of these tests are presented and discussed in this report. From the test results some conclusions are drawn on the possibilities and limitations of pile integrity testing and on the differences in measurements and interpretation methods between the participating companies. Furthermore the validity of the test procedure and the representativeness of the chosen pile shapes for every-day practice is discussed in this report.

3.2 OBJECT OF THE TESTS

Motivation of the selected pile discontinuities for the integrity testing research project. The goal of the research project was to show the possibilities and limitations of integrity testing. For this reason piles with discontinuities were designed and the intention was that these piles would be accessible for many years and for future developments.

For this reason the discontinuities in the piles should be divided into three categories: easy to detect discontinuities, difficult to detect continuities and discontinuities not detectable with the current practice but probably in the near future with improved techniques. The discontinuities were designed with the aid of simulations by a stress wave program.

Testing companies were invited to participate with their integrity testing equipment. The test organizing committee would have considered the design of the discontinuities in he piles unsuccessful when all the participating companies would have been able to detect all discontinuities properly for all piles or when none of the companies would have been able to detect any discontinuity in the piles.

As we expected there was a spread in prediction results and no company was able to predict the discontinuities in all piles. For this reason the design of the discontinuities could be considered successful.

By comparing various interpretation methods and types of equipment a general impression of the accuracy of the available methods and of the spread in results could be obtained.

Apart from these more or less scientific goals, a test competition was held, in which the participating companies could win a price for the best prediction of which pile was which. The result of this competition was considered of secondary interest.

3.3   DESCRIPTION OF THE TEST PROCEDURE

Ten prefab concrete piles with known variations in cross-section and defects were installed at the campus of Delft University in the Netherlands.

Immediately after manufacture of the piles, the pile heads and pile toes were temporarily sealed in the casting yard to make measurements on the piles impossible. After installation of the piles, sealed steel caps were placed over the pile heads for the same reason (Figure 3.1).

Figure 3. 1. Test piles with steel caps.

Each participant received a detailed drawing of each pile and its shape. The ideal of a fully 'blind' test, in which none of the participants has any information on the pile shapes and each participant should describe the pile shape purely on the basis of the measurements, was considered unattainable. It would be impossible for the organising committee to prove that the information on the pile shapes was kept secret.

On the test day, June 16, 1992, the steel caps were opened and the participants were invited to carry out their integrity tests. The tests were performed in a such a way, that the participant did not know which pile was tested.

The measurements were carried out under supervision of the notary public of Delft University of Technology, inside a large tent that was erected around the ten piles. The actual measurements at the pile heads were carried out by an assistant of the notary, out of sight of the participant. This tester was trained on nearby identical driven precast piles prior to testing. The sequence in which the piles were tested was different for each participant and kept secret by the notary. Participants could instruct the tester to re-strike a pile until a good signal was obtained. After completion of all tests, the steel caps were closed again and sealed.

The results, comprising an interpretation of the signals and the assumed sequence in which the piles were measured, were collected by the notary on the 16th of June, 1992.

During the stress wave conference, the company with the best predictions was awarded a price. The results of the other participants were kept secret.

The ten piles will remain available for testing. Furthermore ten piles with the same defects as the test piles in the ground were available above the ground for demonstrations of equipment during the Demonstration Day. Also these piles were covered, with only their heads free accessible. Every visitor of the Demonstration Day was given the opportunity to make his or her own prediction.

3.4   DESCRIPTION OF THE PILES

The ten prefab concrete piles were installed in December 1991. The locations of the piles at the test site at Delft University of Technology are given in Figure 3.2. The shape of the ten test piles is given in Figure 3.3.

SCHOEMAKERSTRAAT

| pile | 6 | 8 | 9 | 3 | 1 | 7 | 4 | 5 | 2 | 10 |
| casing | 1 | 2 | 3 | 4 | 5 | 6 | 7 | 8 | 9 | 10 |

9 x 2.50 m.

PARKING   PLACE

AULA
TECHNICAL UNIVERSITY

Figure 3. 2. Situation of Test piles.

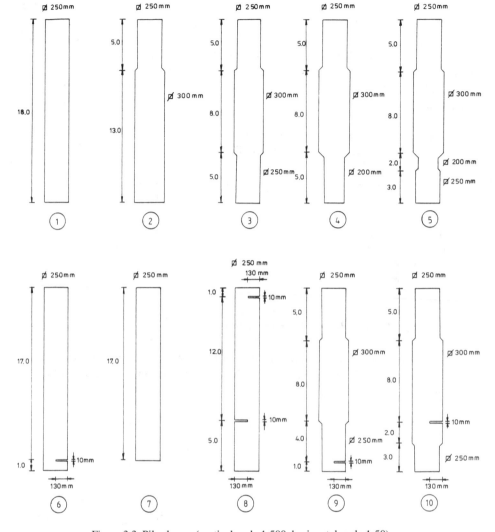

Figure 3.3. Pile shapes (vertical scale 1:500, horizontal scale 1:50).

The pile shapes differ, of course, from shapes met in the daily practice. The differences between the piles were intentionally kept small. It was felt however, that with integrity testing it should be possible, within the limitations of the method, to match the recorded reflectograms with the given pile shapes.

The concrete quality is B55. For this concrete quality an estimated Young's modulus of the piles of 36,000 MN/m^2 is usually adopted. Each pile is reinforced with four rods of 9.3 mm diameter, steel quality: FeB 1860, prestressed to a compression in the concrete shaft of 3.8 MN/m^2.

3.5 DESCRIPTION OF THE SOIL

A detailed description of the soil conditions at the location of the test piles is given in Chapter 2. The soil conditions can be described in general as follows:
- groundlevel at approximately: NAP - 0.9 m.
- groundwater level at approximately NAP - 2.0 m.
- from groundlevel to NAP - 6 m: sandy and clayey layers
- from NAP - 6 m to NAP - 10 m: normally consolidated clay, locally peaty and peat layers
- from NAP -10 to NAP - 14 m: intermediate sandlayer, loose clayey sand
- from NAP - 14 m to NAP - 16 m: clay
- from NAP - 16 m and deeper: dense sands, cone resistances between 10 and 20 MPa.

3.6 INSTALLATION OF THE PILES

Due to the variations in cross-section in the shafts, it was not possible to drive the piles. Therefore an alternative installation procedure, guaranteeing a minimum of risk to pile damage, was followed:

A steel tube ø 570 with a steel shoe of ø 650 at the bottom was driven to the required depth with a IHC-S70 hammer (Figure 3.4). The blow count records are given in Table 3.1 below. The pile numbers correspond with those in Figure 3.2.

After driving a 0.3 to 0.5 m thick layer of sand was placed at the bottom inside the steel tube on which the test piles were placed (Figure 3.5).

The casings were fully backfilled with a bentonite-cement mixture of two different compositions and the driving tube was withdrawn, by upward driving blows (Figure 3.6).

Table 3.1. Blowcount records.

pile tip elevation m - GL	number of blows per 0.25 metre									
	pile 1	pile 2	pile 3	pile 4	pile 5	pile 6	pile 7	pile 8	pile 9	pile 10
15.00 - 15.25	7	17	14	20	20	13	21	-	18	16
15.25 - 15.50	9	13	20	19	21	16	20	19	20	20
15.50 - 15.75	9	10	21	17	21	18	20	18	19	18
15.75 - 16.00	10	22	25	20	21	20	21	18	20	22
16.00 - 16.25	10	28	23	20	20	-	22	19	20	22
16.25 - 16.50	10	20	24	23	23	-	22	24	16	23
16.50 - 16.75	18	23	23	21	21	-	23	22	27	22
16.75 - 17.00	20	25	24	18	26	-	18	21	-	20

Figure 3.4. Installation of the steel casings.

Figure 3.5. Loading of a pile.

Figure 3.6. Adding cement/bentonite.

The composition of the mixture was such that the natural soil conditions were simulated as closely as possible. In the bearing soil layers more cement was added to the mix to obtain a stiffer behaviour compared to the mix surrounding the pile in the soft strata. The influence of the bentonite-cement had been tested before, at an other site. The characteristics of the bentonite-cement mixtures are given in Table 3.2.

Table 3.2. Bentonite-cement mixture.

Depth h	bentonite (kg/m^3 water)	cement content (kg/m^3 water)	ultimate uniaxial strength (kPa)
GL to GL - 15 m	Erbslöh CTN 50 kg	Blast furnace 50 kg	70
GL - 15 to pile tip	Erbslöh CTN 50 kg	Blast furnace 75 kg	210

Immediately after filling with the bentonite-cement, the casing was removed. Sealed steel caps were placed over the pile heads. Figure 3.7 gives a cross-section of the test piles.

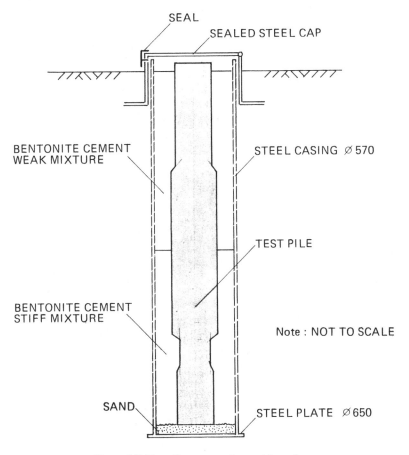

Figure 3.7. Test pile cross-section, not to scale.

3.7   PILE TESTING

3.7.1   Description of applied equipment and interpretation methods

Most companies carried out their own measurements. All participants used the hammer pulse method, in which the pile is struck with a hand held hammer and the pile head velocity or acceleration is measured by a sensor that is placed onto the pilehead. The measurements were presented in the time domain: pile head velocity versus time. CEBTP used an instrumented hammer for their interpretation and presented the results in the frequency domain as well. One of the twelve participants used an analogue system, the other systems were digital.

The participants were somewhat handicapped by the fact that they could not strike the piles themselves: the actual measurements took place inside a tent, out of sight of the participants. Therefore they were given the opportunity to have the pile restruck as often as they liked until a satisfactory signal was obtained.

Not all participants provided information on their interpretation procedures. It is assumed, that the measurements were mostly interpreted visually. Some participants used computer simulations of the measured signals in their analyses. In Table 3.3 the measurement systems and interpretation methods are summarized.

Table 3.3. Measurement systems and interpretation methods.

measuring system	analysis method
CEBTP	not specified: computer analysis gives the pile shape and/or skin friction versus depth.
TNO	visual: depth and nature of reflections analyzed
van Es-Rossmark	visual
IFCO	not specified
GRL	P.I.T. WAP computer simulation
Geomet	visual: depth etc.

3.8   TEST RESULTS

The measured signals for each pile are given in Figures 3.8 through 3.17. In each figure the measured pilehead velocity is plotted versus time. Only 7 different pilehead velocity signals were available for each pile. These were collected with 6 different brands of equipment.

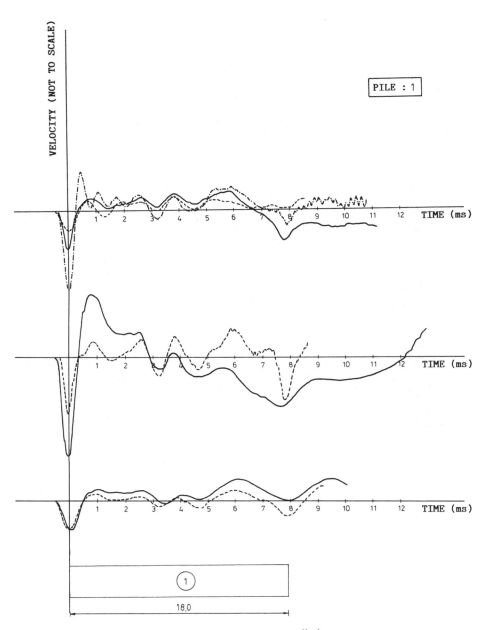

Figure 3.8. Measurements pile 1.

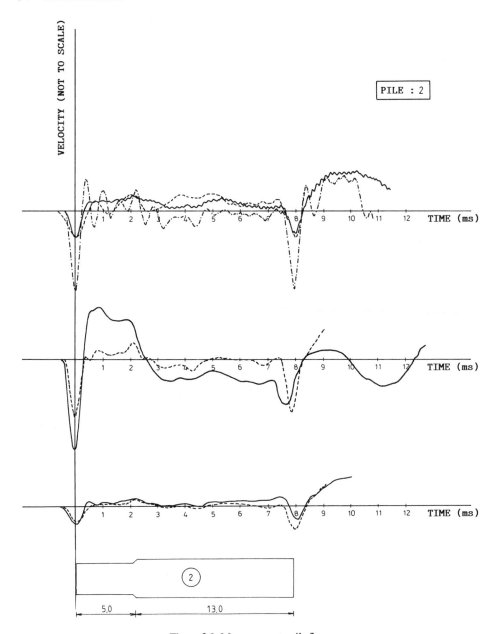

Figure 3.9. Measurements pile 2.

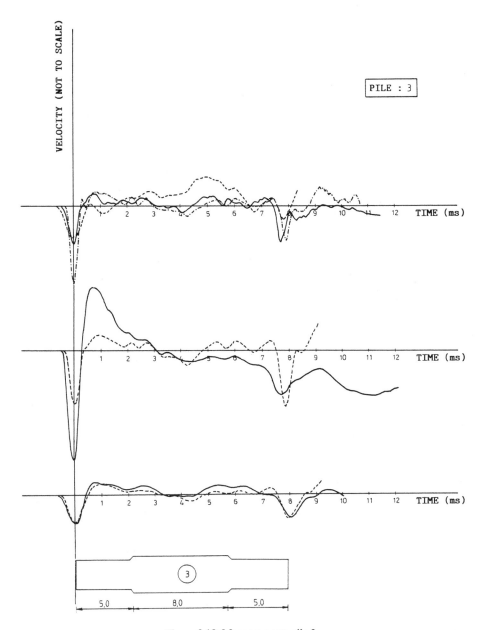

Figure 3.10. Measurements pile 3.

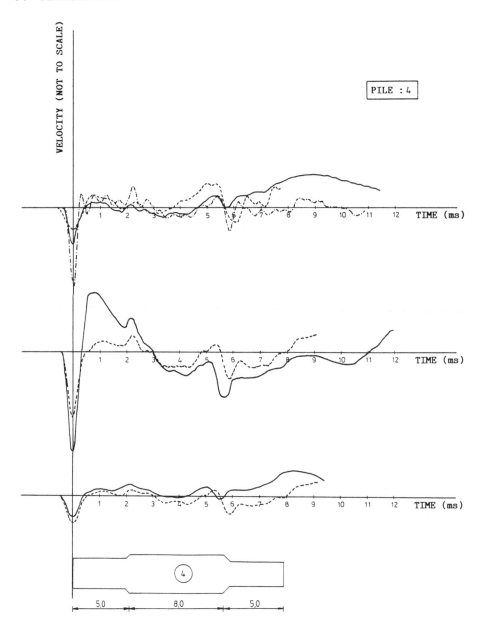

Figure 3.11. Measurements pile 4.

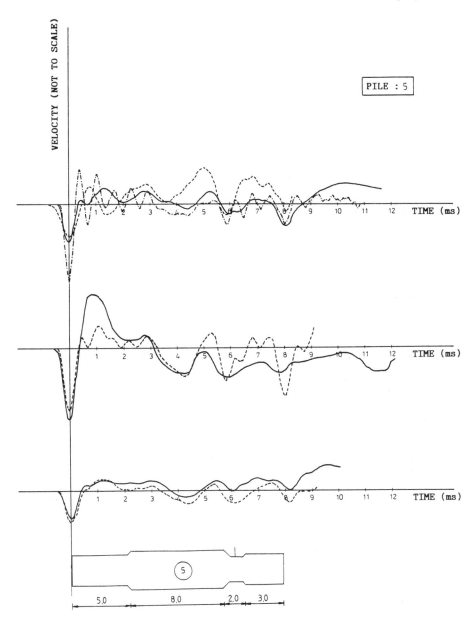

Figure 3.12. Measurements pile 5.

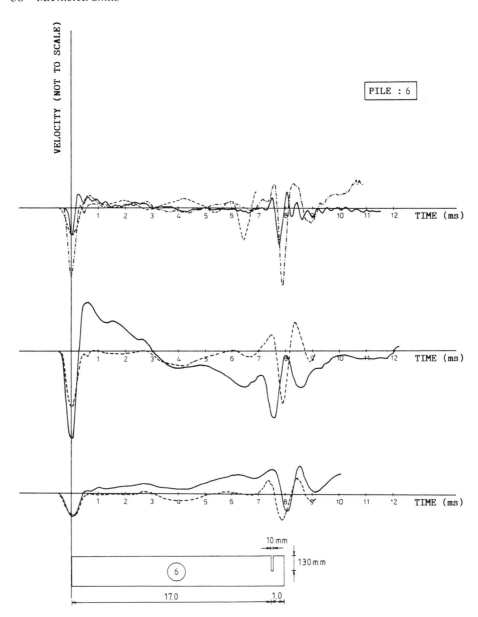

Figure 3.13. Measurements pile 6.

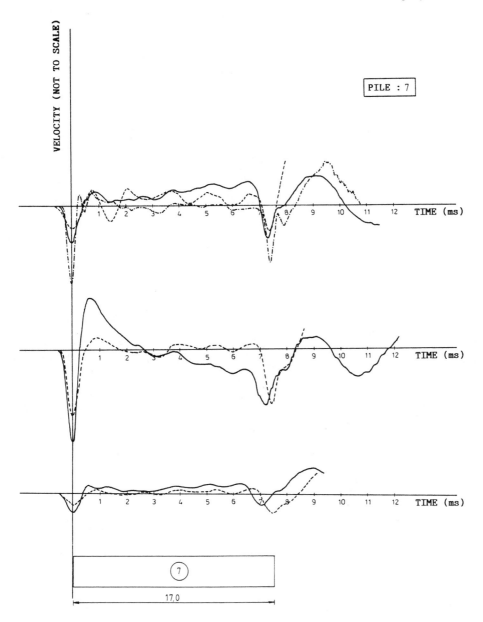

Figure 3.14. Measurements pile 7.

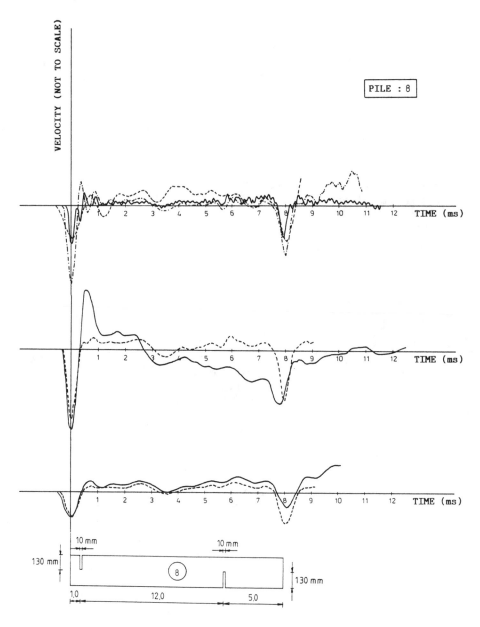

Figure 3.15. Measurements pile 8.

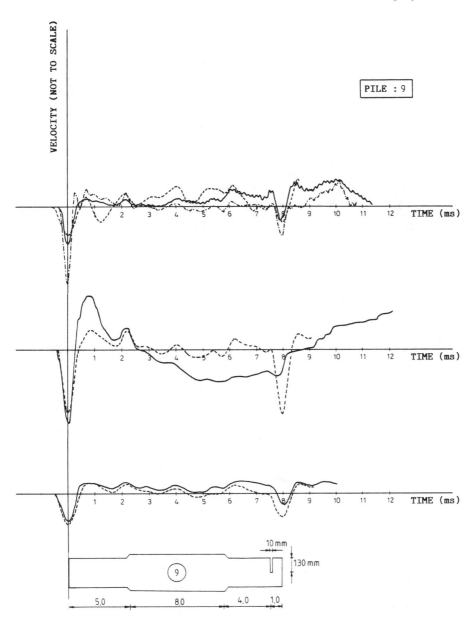

Figure 3.16. Measurements pile 9.

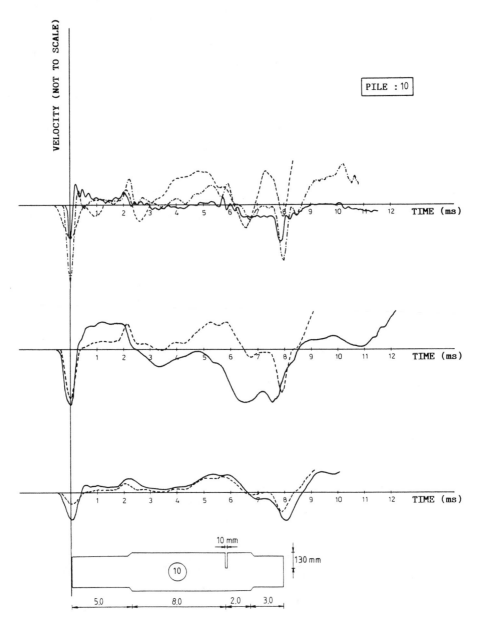

Figure 3.17. Measurements pile 10.

The participants did not plot their results on a time-scale: the pilehead velocity is usually plotted against the pile-length. The conversion from the timescale to the length-scale is as follows:

$$t = 2 \, l/c$$

There are differences in the time-measurements between the various systems. When pile 7, with a known length of 17 metres, is taken as a reference, the 'measured' wave velocities as given by the participants can be computed from the measured pile length and the pre-assumed velocity. As far as information on this wave velocity is given, the corrected velocities were 4737 - 4651 - 4533 - 4665 and 4530 m/s. The maximum difference is about 5%, which is not negligible. It can be concluded that the calibration of the equipment could be improved in this respect.

3.9   TEST COMPETITION

The participants were asked to indicate on the results of the measurements which pile was which. In Figure 3.18 the score per individual participant indicated by the number 1-12 is given. The winner of the test competition was IFCO B.V. from Gouda, the Netherlands. This company identified 7 piles out of the ten piles correctly. The average score was 4.4.

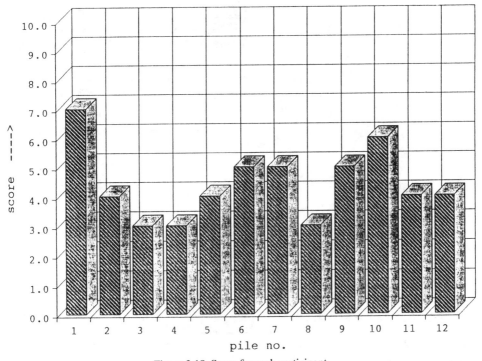

Figure 3.18. Score for each participant.

The numbers of the participants in Figure 3.18 do not correspond with the sequence in which they were listed in Table 3.3: apart from the results of the winner, the results were kept anonymous.

3.10   EVALUATION OF TEST RESULTS

3.10.1   *Results per pile*

In Figure 3.19 the percentage of correct identifications is given for each pile on a scale from 0 to 100%.

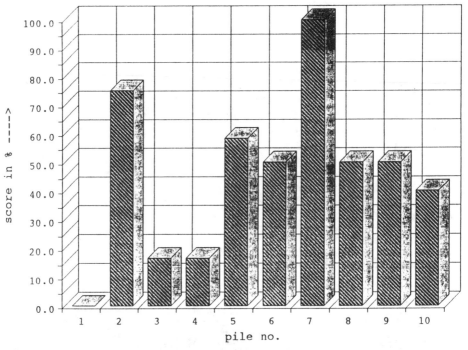

Figure 3.19 Score per pile.

The average score was 44%. The score per pile differs from 0% for pile 1 to 100% for pile 7. Table 3.4 summarizes the results for each pile and the interpretation of each participant. From Table 3.4 it can be concluded that:

Pile 1 was not identified by anyone. This pile was not consistently mixed up with a specific other pile. Pile 2 was mistaken for pile 1 three times. Pile 3 was mistaken for a number of other piles. Only 2 times was pile 3 mistaken for pile 4. It is surprising that this pile is mistaken for pile 8 on three occasions. Pile 4 was mistaken for pile 10 six times. Pile 8 was mistaken for pile 5 three times. Pile 10 was mistaken for pile 4 five times and mistaken for pile 3 three times.

Table 3.4. Results for each participant and each pile.

Participant	Pile number										
	1	2	3	4	5	6	7	8	9	10	S
1	6	C	C	C	C	8	C	1	C	C	7
2	9	C	6	10	C	3	C	C	1	4	4
3	3	C	8	9	10	C	C	5	1	4	3
4	8	C	9	10	3	C	C	1	5	4	3
5	6	C	8	3	4	1	C	5	C	C	4
6	9	C	8	1	C	C	C	4	3	C	5
7	2	1	4	10	C	C	C	C	C	3	5
8	6	C	5	10	9	1	C	C	3	4	3
9	2	1	4	10	C	C	C	C	C	3	5
10	6	C	10	C	C	1	C	C	C	3	6
11	3	C	6	10	C	9	C	C	1	4	4
12	8	1	C	2	4	C	C	6	C	C	5
S	0	9	2	2	7	6	12	6	6	4	

C   = Correct
1 to 10 = Pile was indicated by the participant

3.10.2   *Quality of the measurements*

Nine out of the twelve participants presented their measurements. This resulted in 7 different measurements in the time domain. In Figures 3.8 through 3.17 the measured velocities are plotted against time. It is assumed that most of these velocity signals were obtained by integration of the measured acceleration of the pilehead. The analogue measurement can easily be distinguished by the low frequency component that is visible directly after the input pulse. This distortion, that does not necessarily influence the interpretation, is caused by the analogue integration of the measured acceleration.

Because the participants were handicapped by the fact that they could not strike the piles themselves but manufacturers of sonic pile test equipment claim that reliable results can be obtained, even if inexperienced persons handle the transducers and strike the piles.

From Figures 3.8 through 3.17 it can indeed be concluded, that the measured signals do not differ much. This observation is confirmed by the fact that not one of the participants succeeded to identify pile 1: the measurements all show a distinct reflex at about 3 milliseconds, i.e. at a depth of 6.5-7 metres below the pilehead. Because of the similarity of the results, any inadequacy of the pile tester or the equipment can not explain this reflex. There must therefore be a discontinuity at this depth, either in the shaft or, more likely, in the shaft friction. It is very unfortunate that this discontinuity exists, because pile 1 was intended to be the reference pile.

In some measurements a high frequency disturbance is observed just after the impact: the signals of pile 2 (Figure 3.9) and pile 8 (Figure 3.15) illustrate this. In the case of pile 8 this disturbance may be caused by the 10 mm wide cut in the shaft at 1 metre below the pile head, but for pile 2 there must be a different reason. The disturbances may be caused by the execution of the test. It is remarkable, that the most obvious high frequency disturbances are found in the signals with the shortest impact. One of the participants reported that the interpretation of four out of the ten test results had been seriously hampered by these oscillations. Numerous retests on the testing day did not result in better signals.

For pile 6, one of the signals stops at approximately 7 ms (Figure 3.13).

Apparently the measuring firm mistook the reflex at 6.5 ms for the piletip. This strong reflex can not be explained. It is not unlikely that the magnitude of the reflex is exaggerated by the amplification of the measured signal near the alleged piletip. The reflex mentioned is probably the only essential anomaly that was observed in the measured signals.

The variation in the time scales of the measured signals are in the order of ± 2.5%. The length of the piles and the depth of the discontinuities can therefore be ascertained with an accuracy of 0.5 metre (or better).

3.10.3   Interpretation methods

computer simulations

It is assumed, that the measurements were mostly interpreted by visual methods. Some participants used computer simulations of the measured signals in their analyses however. In Figure 3.20 the results of TNOWAVE and PITWAP computer simulations are given. In most of the TNOWAVE simulations the skin friction was not taken into account.

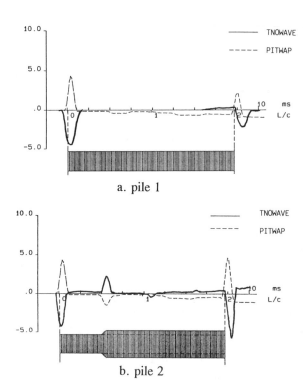

a. pile 1

b. pile 2

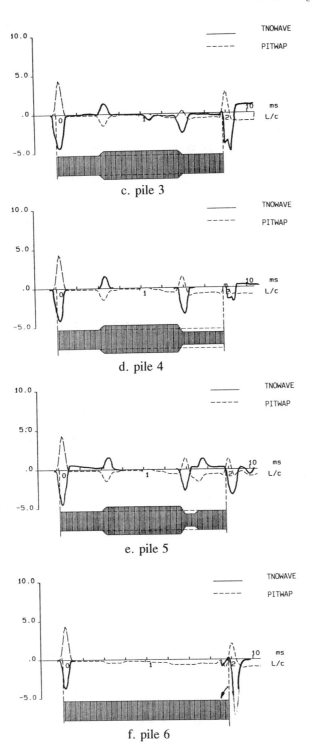

c. pile 3

d. pile 4

e. pile 5

f. pile 6

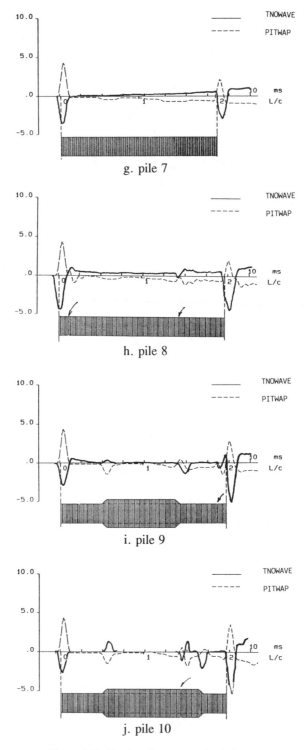

Figure 3.20. Results of computer simulations.

An important aspect of the computer simulations is the answer to the question: can partial cracks or cuts be identified by integrity testing. In the computer simulations the partial cracks are clearly visible.

A partial crack may be interpreted as a reduction in cross-sectional area of the pile with a limited length. Application of the one-dimensional wave-theory then leads to the conclusion, that the reflection from the crack depends on the length of the cross-section reduction (see Rausche et al. 1988). If the length of the reduced cross-section is much smaller than the length of the impact wave the reflection vanishes completely. In this case the length of the impact wave is 2-3 metres whereas the length of the reduced cross-sectional area in piles 6, 8, 9 and 10 is 10 mm. One-dimensional wave theory would therefore indicate that these cracks are invisible.

Indeed the cracks in piles 6, 8, 9 and 10 are barely visible in most of the measured signals. Some of the measurements do give an indication of the crack locations, however. In reality, partial bending cracks in the pileshaft have a width of almost nil and are therefore invisible. Only if the crack runs through the whole cross-section of the pile, the stress wave is totally reflected, but only if the pile section has not been prestressed or reinforced.

The detection of cracks just below the pile head, such as in pile 8, are also dependant on the exact position of the hammer impact on the pile. An experienced tester may detect such a crack by tapping the pilehead at different locations or even by the sound the hammer blow produces. Furthermore the velocity level of the impact gives an indication of cracking close to the pile head.

The visibility of partial cracks in the integrity test results may be better than follows from one-dimensional wave theory because of three-dimensional effects. This is mentioned by Schellingerhout on page 319 of the conference proceedings and confirmed by the tests carried out by Hartung et al. (page 265 of the proceedings). Hartung et al report the results of tests on plastic model piles having similar irregularities to the conference's test piles. Although the length of the irregularities is small compared to the length of the pulse (a ratio of 1 to 30 approximately), the reflection from the 'crack' is in that case clearly visible.

Further research on the reflection amplitudes for cracks as a function of the width of the crack and percentage of the cross-sectional area that is cracked is needed. Furthermore the discrepancy between the computer simulations, in which the partial cracks are clearly visible, and most of the measurements requires further clarification.

Visual interpretation
In Figure 3.21 the test results from one of the participants are given.

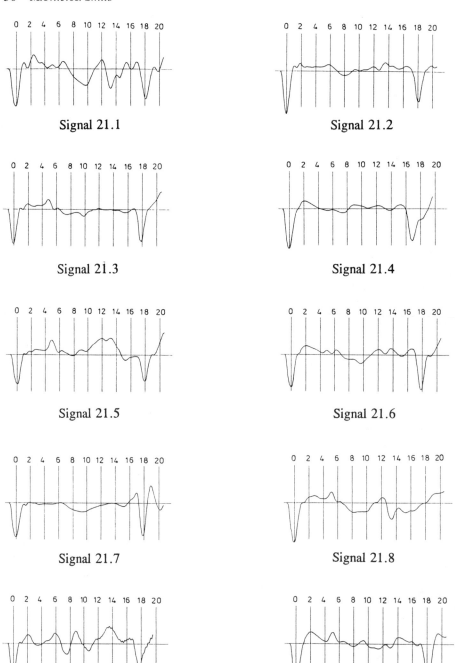

Figure 3.21. Signals of one of the participants.

The procedure for the visual interpretation that most of the participants tried to follow is outlined below.

The pile length measurement is accurate enough to identify the shorter pile, number 7. This is signal number 21.4. Of the remaining 9 piles, the first 13 metres of 6 piles are exactly identical. 3 Piles do not show a sudden impedance increase at 5 metre below the pilehead: piles 1, 6 and 8. Signals 21.2, 21.7 and 21.9 seem to miss reflections at 5 and/or 13 metres, so these signals should correspond to the pile group 1, 6 and 8.

As described above, partial cracks are probably not visible in the measurements. Assuming an equal damping along the shafts of piles 1, 6 and 8 it is therefore reasonable to suppose that the pile having the largest toe reflex is the pile with the lowest number of defects. Following this rule, piles 21.7, 21.2 and 21.9 would be identified as piles 1, 6 and 8 respectively. In reality, piles 1, 6 and 8 were 21.9, 21.7 and 21.2 respectively. This fact may be attributed to differences in shaft friction. As mentioned earlier, the signal of reference pile 1 (21.9) shows a reflex at 7 metre below the pile head which can only be explained by a discontinuity, either in the shaft or in the shaft friction.

In the signal 21.2 (pile 8) a small reflex at 13 metre below the pile head is visible. It should be stated here, that the small reflexes caused by the partial cuts are invisible in the signals of most participants. Furthermore it is fair to say that in every-day circumstances the above mentioned reflection in 21.2 would probably be overlooked and not be associated with cracking.

The remaining 6 piles (2, 3, 4, 5, 9 and 10, Figures 3. 21.1, 3.21.3, 3.21.5, 3.21.6, 3.21.8 and 3.21.10) should all produce more-or-less the same signal in the upper 13 metres. This is not the case: the measurements show disturbances and the amplitude of the reflection at 5 metres below pile head differs strongly. This may be due to differences in shaft friction or to haircracks in the piles that increase the internal damping of the shockwave.

Further evaluation may be carried out on the last part of the signals: for piles 5 and ten reflections at approximately 3 metres above the pile toe should be visible. The signals, 21.1 and 21.5 seem to indicate such reflections. Due to the decrease in cross-section at 5 metres above the pile toe and the increase at 3 metres above the toe, the signal of pile 5 should show a pattern of multiple reflections near the pile toe. In signal 21.1 such a pattern, starting at a depth of 13 metre is indeed visible. Piles 21.1 and 21.5 are therefore recognized as piles 5 and 10 respectively.

Of the 4 remaining piles, 2, 3, 4 and 9, first the pile showing no reflection at a depth of 13 metres is isolated. From the comparison between signals 21.3, 21.6, 21.8 and 21.10 it becomes evident, that signal 21.3 shows no reflex at 13 metres. This should be pile 2. Signal 21.8 is most likely pile 4, because of the large amplitude of the reflex at 13 metres below the pile head. This corresponds very well with the large decrease in cross-section at this depth.

The two remaining piles, 3 and 9, are identical apart from the partial crack in pile 9. If this crack is again assumed to be invisible, it is reasonable to suppose that the signal showing the largest toe reflex is pile 3. 21.6 and 21.10 would therefore be identified as piles 3 and 9 respectively. In reality this is indeed the case.

3.10.4   Sources of inaccuracies in the results

The above described visual interpretation may seem rather straightforward, but in reality the interpretation of the results has not been easy. The following tricky circumstances of the tests are believed to be an important source of inaccuracy:

The similarity between the piles. In fact the first 13 metres of piles 2, 3, 4, 5, 9 and 10 are identical. The differences between these piles appear at a depth of about 50 times the pile diameter, which in stiffer soils is usually too deep for any integrity test method.

The width of the 'half-cracks'. According to one-dimensional stress-wave theory, decreases in cross-section of a limited length are not visible. Piles 3 and 9 and piles 1, 6 and 8 would therefore be inseparable. Some measurements seem to contradict this point however.

The variations in cross-section of the piles seem relatively small, especially if the very slender piles are drawn to true scale. The sudden changes in cross-sectional area and therefore in the pile impedance vary however between 30 and 225%, which is substantial.

The influence of the bentonite-cement mixture surrounding the pile shafts. There is enough evidence that the shaft friction differed between the individual piles. This inhomogeneity is not unlike every-day circumstances.

The measurements of pile 1 all show a distinct reflex at about 7 metres below the pilehead, caused by a discontinuity at this depth, either in the shaft or, more likely, in the shaft friction. It is very unfortunate that this discontinuity exists, because pile 1 was intended to be the reference pile.

The lack of experience of the persons who carried out the tests. Since integrity test equipment is sold world-wide on a large scale, this operator dependency would have an adverse effect on the reliability of the test method. This point is partially contradicted by the resemblance of the measured signals. The fact that all participants missed pile 1 and all participants identified pile 7 leads to the conclusion that the influence of the tester should not be exaggerated.

In spite of this in some cases poor quality measurements were obtained. Criticism was vented by some on the integrity tests in this conference because the anomalies in the piles were considered undetectable (see the discussion in Ground Engineering 1993). In view of the complications listed above, the average score of 44% should be seen in a different perspective: it proved integrity testing to be more accurate than perhaps expected but far from sufficient.

Partial bending cracks and horizontal total cracks can be controlled by repeating the tests under pileload conditions. If the reflections disappear, the pile bearing capacity may not be reduced by the cracks.

3.11   CONCLUSIONS

'Hammer pulse' integrity tests were carried out on ten piles by twelve companies. The test methods, the interpretation and the measurements obtained did not differ strongly.

The shapes of the ten test piles were quite similar. Furthermore the quality of the measurements might have been influenced by the lack of experience of the persons who carried out the tests. Due to unexpected reflections, the signal of pile 1 could not be used as a reference, which complicated the interpretation. In view of this the average score of 44% does not disqualify integrity testing as a reliable test technique.

All the defects in the test piles consisted of sudden changes in cross-section. The visibility of a gradually changing cross-section and of differences in concrete quality was not investigated. In this respect the chosen pile shapes were not representative of every-day practice but were easier.

The quality of the measurements can be influenced by the test operator. Experience with holding the sensor to the pile head and tapping the pile with a hammer is beneficial for the quality of the measurements. A frequent replacement of the contact paste is essential when dusty or otherwise unclean pile heads are encountered.

The capability of pile integrity testing to detect sudden increases or decreases in cross-section is better than assumed by some (see discussion in Ground Engineering, 1993). For example the piles three and four were definitely not indistinguishable from each other, the greater cross-section reduction in pile four was clearly visible in the results.

The variation in the time scales of the measured signals were in the order of ± 2.5%. The length of the piles and the depth of the discontinuities could therefore be ascertained with an accuracy of plus minus 0.5 metre. This small variation is the result of a constant concrete quality obtained by using prefabricated piles.

The measurements of some of the participating companies show small reflections at the locations of the cuts in piles 6, 8, 9 and 10 that can not be totally coincidental. These cuts are also apparent in computer simulations. In everyday practice these small reflections would go by unnoticed, as variations in soil profile may give similar reflections.

Integrity testing in practice is performed to detect major defects that endanger the bearing function of the pile and not to detect differences between rather similar piles. Therefore the average score of 44% in this complicated competition does not prove that integrity testing is an unreliable test method for its purpose. If the user is well aware of the limitations of the method and uses other sources of information in the interpretation as well, such as details about the pile installation and soil conditions, integrity testing is a powerful and cost-effective tool to check the quality of piles after installation. In this context it should be stated, that most participants to the test competition were able to submit a preliminary interpretation of the measurements within 24 hours after testing. None of these participants changed their preliminary conclusions on the basis of closer examination and analyses during the weeks thereafter.

TNO has performed additional measurements in March 1993 because it became clear to TNO that during the original tests in September 1992 the connection between the transducer and the pile was defective in a number of tests. Better results were then obtained.

3.12   RECOMMENDATIONS

The results of pile integrity testing largely depend on the type and quality of the equipment used and on the interpretation by the consultant. Certification and standardisation of equipment, of presentation of the results and of interpretation is desirable.

The tests have to be performed very carefully; small irregularities in the testing procedure may result in important errors.

In view of possible misjudgements consultants should be very restrained especially when the interpretation lead to disapproval of piles. If possible a visual control should be performed via excavation of the pile shaft.

Further research on the reflection amplitudes for cracks as a function of the width of the crack and percentage of the cross-sectional area that is cracked is needed. Furthermore the discrepancy between the computer simulations, in which the partial cracks are clearly visible, and most of the measurements requires further clarification.

REFERENCES

Rausche, F, Likins, G.E. and Hussein, M. Pile integrity by low and high strain impacts Proc. of the third international conf. on the application of stress wave theory to piles, Ottawa 44-55
Ground Engineering February, April and June 1993:
Stain, R. Test's Integrity is questionable *(February, p. 7)*
Weele, A.F. van Integrity Test was not questionable *(April; p. 15)*
Stain, R. Competition was not applicable *(April; p. 15)*
Ellway, K. Objectives of competition are unclear *(June; p. 8)*

CHAPTER 4

Prediction of load-displacements characteristics of piles from the results of dynamic/kinetic load tests

J. GEERLING
Delft Geotechnics, Delft, Netherlands

M.Th.J.H. SMITS
Fugro BV, Leidschendam, Netherlands

ABSTRACT: Dynamic and kinetic pile load tests were carried out on four out of five precast concrete piles at the test site. For verification of the predicted bearing capacity and load-displacement behaviour the five piles were subjected to a static load test.

The primary goal of the tests was to gather information in order to improve the insight in the current state-of-the-art regarding dynamic/kinetic load testing of piles.

The tests also served as a kind of a competition. The eight participating companies could get a reward for the best prediction of the maximum bearing capacity and the load deformation behaviour.

Special measurements have been carried out on the piles and in the ground to enhance the understanding of the phenomena which occur during driving and dynamic testing. If any drawbacks of dynamic load testing methods would emerge, the results of these additional measurements may help to find the reasons for these drawbacks and perhaps can be valuable for the improvement of the analysis which relates the dynamic to the static behaviour. By comparing various interpretation methods and types of equipment, a general impression regarding the accuracy and the spreading of the various dynamic/kinetic methods has been obtained.

Quite a large scatter has been observed between the predictions of the various competitors.

The following participants gave the best prediction in the respective categories:
– For the prediction of pile head displacement; Goble & Rausche Likins and Associates Inc, Ohio, USA;
– For the prediction of the maximum bearing capacity; Fugro McClelland Engineers bv, Leidschendam, The Netherlands;
– For the prediction of the overall behaviour; Pile Dynamics Europe, Vastra Frolunder, Sweden.

4.1 INSTALLATION OF THE TEST PILES

The five precast concrete piles were driven on June 12, 1992. During driving the following measurements were carried out:
– Acceleration and strain in the piles;
– Vibration measurements in the surroundings;
– The number of blows per meter over the full length of the piles.

Table 4.1. Pile lengths and levels after pile driving and before the dynamic pile load tests.

Pile no.	Pile length in m	Pile tip level in m below NAP	Pile head level in m below/above NAP	Groundlevel related to NAP (m)
1	11,39	−11,56	−0,17	−0,83
2	14,92	−14,90	+0,02	−0,81
3	18,96	−19,07	−0,11	−0,72
4	18,95	−19,11	−0,16	−0,88
5	18,93	−19,02	−0,09	−1,03

Table 4.2. Pile length and levels before the static pile load tests.

Pile no.	Pile length in m	Pile tip level in m below NAP	Pile head level in m below/above NAP
1	11,50	- 11,89	- 0,84
2	14,69	- 14,91	- 0,22
3	19,08	- 19,23	- 0,15
4	18,98	- 19,11	- 0,13
5	18,96	- 19,06	- 0,10

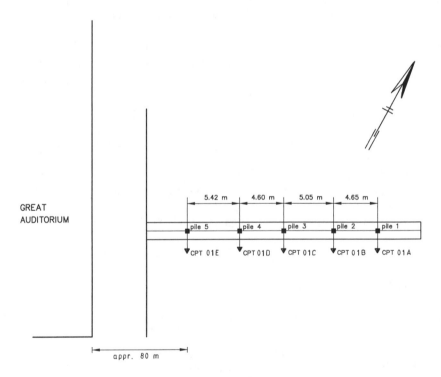

Figure 4.1. Pile lay-out.

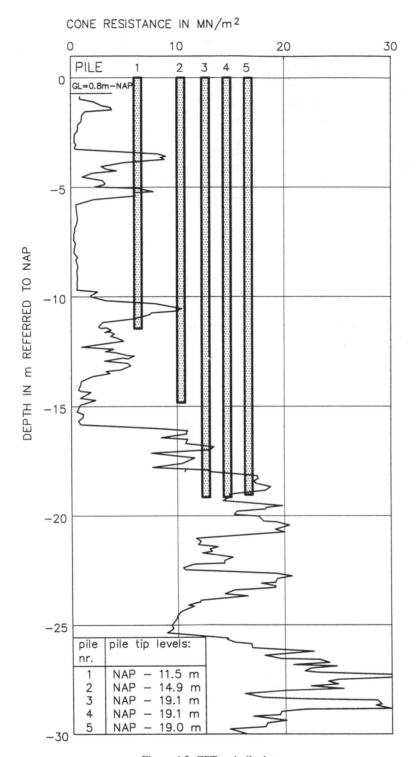

Figure 4.2. CPT and pile data.

In this section only the driving results are presented. The results of the accelaration, strain and vibration measurements are discused in Section 4.4.

The test piles 1 to 5 were precast concrete piles, square 0,25 m, long 11 to 19 m. They were manufactored on May 13, 1992. The piles were placed in a row, with a centre-to-centre distance of approximately 5 m.

In Table 4.1 the pile length, pile tip levels and pile head levels are summarized, after the pile installation was finished, but before the dynamic load tests (June 15, 1992).

Due to the dynamic pile load testing the head and tip levels changed as can be concluded from Table 4.2. In this table pile length, pile tip and the head levels just before the start of the static load tests on June 28, 1992 are summarized.

Some pile heads were broken during driving and had to be repaired. This caused a lengthening of some of the piles.

The quality of the concrete of the piles was B55. According to the manufacturer, the estimated Young's modulus of the piles is 36 000 MN/m^2. Each pile is pre-stressed to 4,8 MN/m^2 by 6 strands of 9,3 mm diameter, steel quality: Feb 1860.

The locations of the piles at the test site behind the Great Auditorium of the Delft University of Technology are shown on the sketch of the lay-out of the testsite in Figure 4.1.

The pile tip elevations with respect to the cone resistance diagram of CPT 01D are shown in Figure 4.2.

It shows the following:
– Tip of pile 1 is in the intermediate sand layer (NAP –11,56 m);
– Tip of pile 2 is in a clay layer (NAP –14,90 m);
– Tip of pile 3 is in the dense sand layer (NAP –19,07 m);
– Tip of pile 4 is in the dense sand layer (NAP –19,11 m);
– Tip of pile 5 is in the dense sand layer (NAP –19,02 m).

The piles were driven with the hydrohammer type ICE SC-40. Table 4.3 gives the data of this hammer.

The blowcount records are summarized in Tables 4.4a-4.4e.

Table 4.3. Specifications ICE SC-40 hammer.

ICE SC-40 Hammer	
Energy range (kNm)	4-38
Blows per minute	± 40
Ram mass (kg)	2 550
Maximum stroke (m)	1,5

Table 4.4a. Blowcount record of pile 1 (GL = NAP – 0,83 m).

Pile tip penetration from ... to ... m – NAP	Number of blows per 0,25 m	Energy level (kNm) per blow
8,83 to 9,83	5, 5, 5, 12	2
9,83 to 10,83	10, 23, 50, 27	4
10,83 to 11,537	24, 16, 12	6
Final pile tip penetration NAP – 11,53 m		

Table 4.4b. Blowcount record of pile 2 (GL = NAP – 0,81 m).

Pile tip penetration from ... to ... m – NAP	Number of blows per 0,25 m	Energy level (kNm) per blow
8,81 to 9,81	23, 23	5
9,81 to 10,81	5, 17, 24, 45	6
10,81 to 11,81	19, 26, 24, 22	6
11,81 to 12,81	19, 19, 15, 21	6
12,81 to 13,81	35, 38, 27, 22	5
13,81 to 14,81	28, 27, 27, 30	4
Final pile tip penetration NAP – 14,81 m		

Table 4.4c. Blowcount record of pile 3 (GL = NAP – 0,72 m).

Pile tip penetration from ... to ... m – NAP	Number of blows per 0,25 m	Energy level (kNm) per blow
9,72 to 10,72	12, 12, 20, 23	6
10,72 to 11,72	14, 8, 14, 14	6
11,72 to 12,72	13, 10, 12, 17	6
12,72 to 13,72	25, 22, 22, 10	5
13,72 to 14,72	9, 8, 9, 12	5
14,72 to 15,72	10, 6, 7, 18	4
15,72 to 16,72	45, 40, 40, 27	3/10
16,72 to 17,72	27, 23, 19, 13	16
17.72 to 18,72	17, 14, 14, 16	28
18,72 to 19,12	15	28
Final pile tip elevation 1 NAP – 19,12 m		

Table 4.4d. Blowcount record of pile 4 (GL = NAP – 0,88 m).

Pile tip penetration from ... to ... m – NAP	Number of blows per 0,25 m	Energy level (kNm) per blow
10,38 to 10,88	14, 10	7
10,88 to 11,88	7, 7, 7, 6	13
11,88 to 12,88	6, 6, 5, 7	14
12,88 to 13,88	7, 15, 7, 6	14
13,88 to 14,88	6, 5, 6, 3	16
14,88 to 15,88	3, 4, 4, 8	16
15,88 to 16,88	10, 20, 21, 26	16
16,88 to 17,88	24, 21, 17, 15	22
17,88 to 18,88	14, 13, 18, 18	31
18,88 to 19,08	15	31
Final pile tip penetration NAP – 19,08 m		

Table 4.4e. Blowcount record of pile 5 (GL = NAP − 1.03 m).

Pile tip penetration from ... to ... m − NAP	Number of blows per 0,25 m	Energy level (kNm) per blow
10,03 to 11,03	4, 12, 13, 11	10
11,03 to 12,03	7, 4, 4, 6	10
12,03 to 13,03	5, 3, 7, 8	11
13,03 to 14,03	12, 7, 8, 3	9
14,03 to 15,03	3, 3, 3, 3	10
15,03 to 16,03	2, 1, 3, 3	10
16,03 to 17,03	17, 27, 27, 28	11
17,03 to 18,03	31, 34, 37, 35	11
18,03 to 19,03	37, 29, 32, 41	14
Final pile tip penetration NAP − 19,03 m		

The variation in blowcount between the piles at various depths is mainly due to the variation in energy per blow.

The soil profile and soil strength in the area, where the piles were placed, can be considered as uniform.

In pile nr 3 also strains and accelerations were measured at several levels in the shaft during driving. The results are discussed in Section 4.4, together with the results of measured accelerations, porewater pressures and soil pressures at a depth of 18 m below groundlevel in the vicinity of this pile.

4.2 DESCRIPTION OF THE DYNAMIC AND KINETIC PILE LOAD TESTS

On June 15, 1992 the dynamic pile load tests on the piles 1, 2, 3 and 5 were carried out (pile 4 was used as a reference pile and was only statically loaded).

These piles were loaded with the same hydrohammer as used for the installation of the piles.The ram with a mass of 2550 kg, was applied as a drop weight. It was dropped from different heights on each pile in order to subject the piles to impact forces of various energy levels. On the basis of the measurements during these blows, predictions were made of the static load-displacement behaviour of each pile. The predictions were sent to the notary of Delft University.

Predictions were not only supplied by the companies that carried out the measurements, but also by companies and organisations that used the measured signals of others for their estimations.

On June 19, 23 and 24 the kinetic load tests were carried out on the same four piles. In these tests, a relatively long duration impact is produced on the pile head, either by explosives ('statnamic') or by a heavy drop weight with a set of large springs ('quasi-static') attached to its bottom. This results in rather low accelerations of the tested piles.

The predictions of the static load-displacement behaviour of the piles based on the results of the kinetic tests were also submitted to the notary.

Pile 4 was not dynamically or kinematically load tested. It was kept as a virgin pile to investigate the possible influence of the dynamic load tests on the results of the static load tests.

Table 4.5. Dynamic load test: blows on each pile

Pile number	Net energy level (kNm)	Maximum impact force (kN)	Final set per blow in mm
1: first blow	4,0	26	10,6
second blow	4,0	26	10,3
thirth blow	2,0	13	1,0
2: first blow	8,0	52	7,1
3: first blow	23,9	159	8,1
second blow	23,9	159	9,2
thirth blow	23,9	159	9,6
5: first blow	2,0	13	0,4
second blow	1,9	79	0,4
thirth blow	1,9	19	0,5
fourth blow	2,9	159	1,9
fifth blow	2,9	159	2,6
sixth blow	23,9	159	6,0
seventh blow	23,9	159	4,0

Table 4.6. Measuring systems and interpretation methods.

Measuring system	Interpretation method
Fugro	CAPWAP
PDE	CAPWAP
IFCO	Lumped mass
–	Cone Penetration Tests
TNO	Lumped mass
PDE	CAPWAP
TNO	TNOWAVE
TNO	TEPWAP and energy approach
TNO	NUSWAP and CAPWAP

4.2.1 *Dynamic tests*

In Table 4.5 for each pile.the net energy level, the maximum impact force and the final set per blow are summarized in chronological order.

The participants were invited to install their sensors on the head of the testpiles. Figure 4.3 shows a photograph of a pile head just before the dynamic test.

The following participants carried out the indicated measurements:
– Accelerations and strains (FPDS system): TNO;
– Displacements and strains: IFCO;
– Accelerations and strains: Fugro;
– Accelerations and strains (PDA system): Pile Dynamics Europe.

The other participants based their predictions on these measurements.

Most participants have given very limited information regarding the methods used for the interpretation of the results of the dynamic tests.

In processing the data and the conversion into static pile behaviour, a variety of methods was used. More details about the elaboration of the results are presented in Sections 4.6.1-4.6.3.

Table 4.6 gives an overview of the measuring system and interpretation method

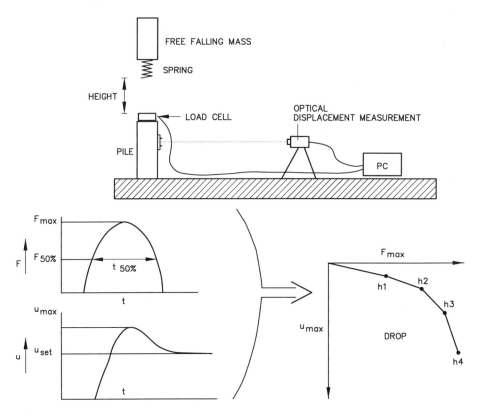

Figure 4.3. Details about the pseudo-static load test.

used by the participants. It appears that the measuring system and interpretation methods are interchangeable.

For the sake of anonymity the participants in the contest are in this book referred to by capital A-K.

4.2.2 *Kinetic tests*

In kinetic testing (quasi static or statnamic) the deceleration and acceleration of a falling mass is used to develop an inertial force of relatively long duration on the head of the pile. During the blow the pile head force and and the movement of the top of the pile are measured. The duration of the pulse is in the order of 100 to 200 milliseconds. This is long enough for the stress waves to travel back and forth several times within the pile. As a result of the long duration and the gradual build-up of the force pulse, inertial effects are considered to be of minor importance.

Details of the kinetic testing method and the comparison with other methods of dynamic testing are represented in the keynote lecture presented in this conference by Holeyman (1992).

A 'quasi-static' pile load test is carried out by a heavy mass of 25 tons which drops from a certain height onto the pile. A set of coil springs is attached to the bottom of the weight. This results in an extension of the duration of the impact force.

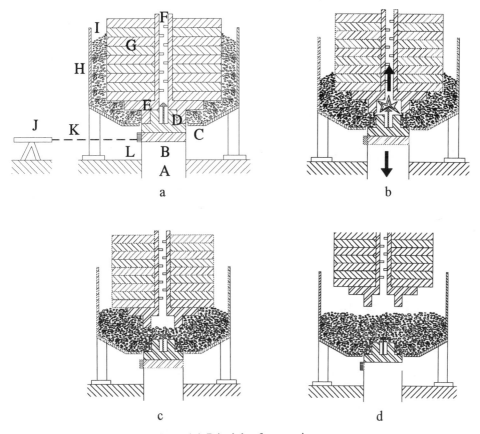

Figure 4.4. Principle of statnamic test.

The mass is moved and lifted by a mobile rig to enable testing of several piles per day. During the test the pile head force and the pile head displacement are recorded. In Figure 4.4 the scheme of the test set-up is shown. It also gives an example of the load versus time and the displacement versus time diagrams. More details on this test method were described by Gonin et al. (1984).

The quasi-static tests were carried out on June 19. In Table 4.7 some characteristics of the blows on pile number 3 are given.

A 'statnamic' pile load test is carried out by a controlled fuel explosion between the pile head and a heavy mass on top of the pile. When the explosion takes place, the reactive mass is pushed upwards, and the pile is pushed downwards. This process yields a relatively long duration impact on the pile head. In Figure 4.4 an impression of the test set-up is given. Details on this test method are described by Bermingham & Janes (1989) and by Middendorp et al. (1992).

The reactive mass is embedded in gravel which prevents the mass from falling back on the pile head after the explosion. The load is measured by means of a load cell; the pile head displacement by a laser sensor. The test method can be used for very high capacities by manipulating the reactive mass, the properties of the fuel in the pressure chamber and the characteristics of the pressure chamber.

Table 4.7. Results of quasi-static load tests on pile number 3.

Height H (m)	maximum impact force (MN)	duration of the pulse $t_{50\%}$ (ms)	final set in mm
0,20	1,20	114	0,4
0,50	1,65	113	2,5
0,75	1,94	111	5,2
0,85	2,05	110	8,8

Table 4.8. Statnamic load tests: blows on each pile

Pile number	Time	Maximum impact force (kN)	Duration of the pulse (ms)	Final set in mm
1:	23/6/92 22.45 h	1450	140	220
2:	24/6/92 15.00 h	1090	140	5
3:	24/6/92 21.00 h	1800	140	9
5:	23/6/92 15.00 h	2190	140	16

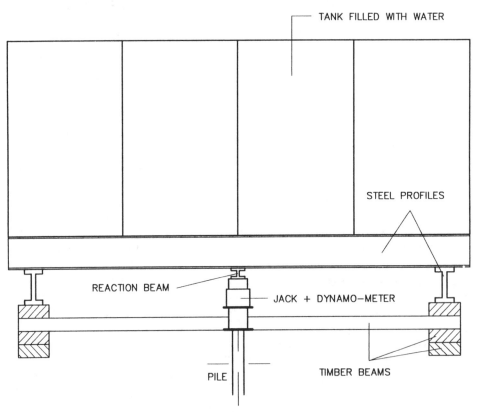

Figure 4.5. Set-up for static load test.

The statnamic tests were carried out on June 23 and 24. Table 4.8 gives an overview of the test results in terms of maximum impact force and duration of the pulse and final set of the piles.

4.3 STATIC PILE LOAD TESTS

As a reference for the predictions, the test piles 1 to 5 inclusive were subjected to static pile load tests. The static pile load tests were carried out in the period between June 29 and July 8, 1992. After completion of the static load tests, the results of the predictions of the participating firms were compared with the static load test results.

The static load tests were carried out in accordance with: 'ISSMFE Subcommittee on Field and Laboratory Testing', 'Axial pile loading test-Part 1: Static Loading'.

The reaction force for the loading tests was supplied by dead weight consisting of four tanks filled with water. Each full tank supplying 500 kN. Figure 4.5 shows the system with the water tanks and the loading system on the piles. In order to keep the movement of the reaction system under control the vertical displacements of the reaction system were measured periodically during the pile load testing.

The loading scheme is shown in Figure 4.7. The load has been applied in steps of 0,125 × the estimated maximum bearing capacity. Each load step lasted 1 hour. After the 4th step the load has been removed from the test pile in steps. A small initial load of 10 kN was kept on the pile head for about half an hour. Thereafter the pile has been reloaded until failure. The duration of each load step in the unload-reload loop was 10 minutes.

Loading was continued until a displacement of the pile head of 150 mm. This concurs with approximately 50 % of the pile diameter. Then the pile was unloaded.

The load on the pile head was applied by a hydraulic jack, with a capacity: of 5000 kN and a maximum stroke of 212 mm.

The loads during the various loading steps were kept constant with an accuracy of 0,5% of the maximum load by means of an electronic controll system.

The hydraulic pump was capable of increasing the load with approximately 10 kN per second.

The following measurements were carried out:

– The load on the pile head: measured by a 2000 kN capacity dynamometer, located between the hydraulic jack and a the main reaction beam of the kentledge installation. The sample time was 5 seconds.

– The pile head displacements: measured with a levelling instrument and simultaneously with a laser beam device. Three rulers were connected to the pile head to measure the vertical pile head displacement at three points. A longline diode was connected to the pile head for laser beam measurements in vertical direction. A nearby pile was used as a fixed reference level. The laser beam measurements were recorded at 5 second intervals on a data logger. The measurments with the levelling instrument were taken at intervals of 2 to 20 minutes.

– The displacement of the pile tip: measured by means of a tell tale. This device consisted of a free standing steel rod in a pipe which was placed in the axis of the pile during manufactoring. The displacements of the tell tale were measured with the levelling instrument. They were also determined at 5 second intervals, by an electri-

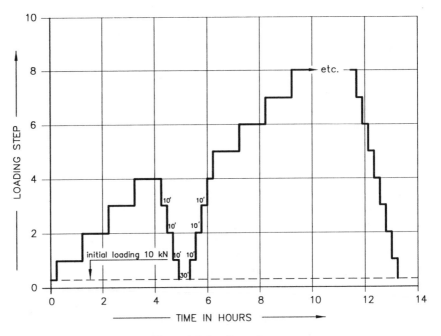

Figure 4.6. Loading scheme.

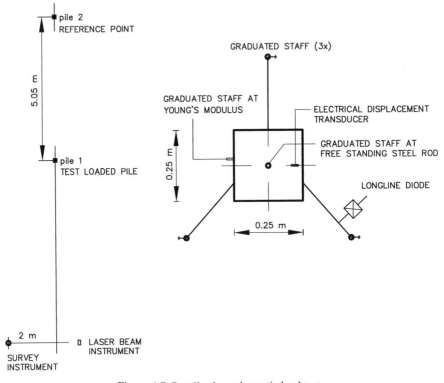

Figure 4.7. Details about the static load tests.

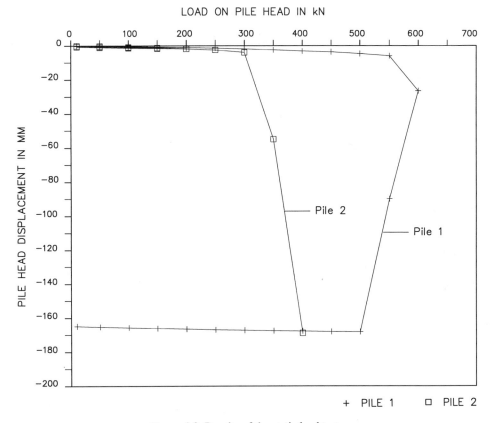

Figure 4.8. Results of the static load tests.

cal displacement transducer, which measured the displacement between top of the tell tale and the pile head.

– Young's modulus of the pile material: measured by means of a 239 mm long strain gauge, fixed to the concrete shaft of the pile 0,5 m below pile head.

Particulars about the placing of the various instruments for the measurements are shown in Figure 4.8.

The following notes concern the most important events which occurred during the static load tests on the test piles 1 through 5. Apart from some minor divergences the tests were carried out in accordance with the standard test procedure.

Pile 1
The test on this pile started on June 29, 1992.

The pile was loaded in steps of 50 kN; the 'initial' loading step was 10 kN.

During the application of the 8th load step the pile head displacement exceeded the required minimum value of 150 mm. The load could not be increased anymore and the test had to be stopped. The pile was unloaded without taking any readings.

The strain gauge for the measurements of Young's modulus could not be installed because of lack of space. The weather was hot and sunny.Temperature reached 28° centigrade.

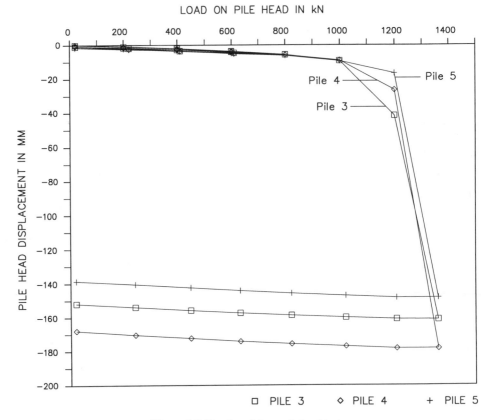

Figure 4.9. Results of the static load tests.

Pile 2

The test on this pile started on June 30, 1992.

The load on the pile was increased in steps of 50 kN; the 'initial' loading step was 10 kN. Approximately 30 minutes after the application of the 12th load step, the pile head displacement exceeded the required minimum value of 150 mm. The load could not be increased further and the load was taken away.

During the first 2 hours the weather was hot, sunny and windy. Later the temperature dropped and the sky became cloudy.

Pile 3

The test started on July 2 1992.

The load on the pile was increased in steps of 200 kN; the 'initial' loading step was 20 kN.

During the application of the 7th load step, the pile head displacement exceeded 150 mm. So the load was removed.

The weather conditions were: cloudy and rainy.

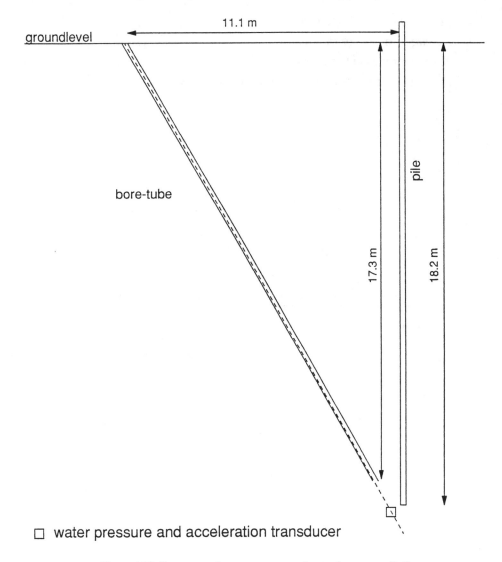

Figure 4.10. Test set-up for waterpressure observations near pile 3.

Pile 4

The test started on July 6, 1992.

The load on the pile was increased in steps of 200 kN; the 'initial' loading step was 20 kN.

During the load step of 1400 kN, the pile head displacement exceeded the required minimum of 150 mm. Then the pile was unloaded.

The weather was sunny and windy.

Pile 5

The load test started on July 8, 1992.

The pile was loaded in steps of 200 kN: the 'initial' loading was 20 kN.

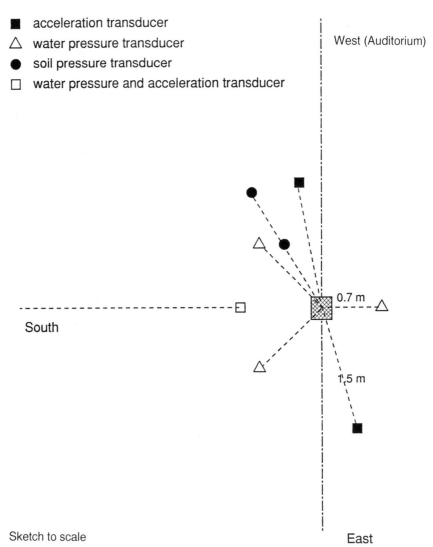

■ acceleration transducer
△ water pressure transducer
● soil pressure transducer
□ water pressure and acceleration transducer

West (Auditorium)

0.7 m

South

1,5 m

Sketch to scale

East

Figure 4.11. Lay-out of sensors near the fort of pile 3.

Approximately 10 minutes after the application of the load of 1400 kN, the pile head displacement exceeded the limiting value of 150 mm. Directly thereafter the pile was unloaded.

The weather was sunny.

Remark: Weather conditions most likely had only minor influences on the measurements.

The results of the static pile load tests are described and presented in detail in the Report of Delft Geotechnics (Delft Geotechnics 1992).

For the estimation of Young's modulus of the pile material the average of the strain gauge measurements of the last 3 loading steps were used. In this way it has been established that Young's modulus for all piles was 30 000 MN/m^2 in stead of

36 000 MN/m^2 as estimated by the manufactorer of the piles.

The load-displacement curve of the pile head as predicted by the participants in the contest on the basis of the dynamic and kinimatic load tests has been compared with the load-displacements curves derived from the measured displacements at the end of the various loading steps of the static load tests.

Figures 4.8 and 4.9 show the load-displacements diagram of the testpiles 1-5.

4.4 SPECIAL TESTS

4.4.1 *Measurements in the soil at 18 m depth*

Normally only measurements at groundlevel or measurements on the pile itself are carried out during piledriving or dynamic testing. To get reliable information on soil response, it is necessary to make measurements in the soil near the pile toe. There-fore, Delft Geotechnics set up a programme to measure acceleration, water pressure and soil pressure at a depth of about 18 m below groundlevel near pile 3. The measurements were executed during the driving of the pile, during dynamic loading tests, during statnamic loading tests and during the static loading tests.

Instrumentation
The instruments were placed in the soil around pile 3. At the top and the toe of this pile acceleration transducers were placed. The pile length was 19 m, the final level of the pile head was 0,8 m above groundlevel.

Four water pressure transducers and nine acceleration transducers were installed in the ground at a depth of 16 to 18 m below groundlevel. The acceleration transducers were installed in three groups, one for the vertical direction and two for the horizontal directions, parallel and prependicular to the direction to the pile. One group of three acceleration transducers and a water pressure transducer were combined to be able to measure the accelerations and water pressure at the same location.

The placed instruments were:
– Three water pressure transducers, at a distance of 1 m from the centre of the pile and at a depth of respectively 17, 17 and 16 m below groundlevel;
– Two groups of three prependicular acceleration transducers at a distance of 1,5 m from the centre of the pile and at a depth of 17 m below groundlevel;
– A combined group of three prependicular acceleration transducers and the water pressure transducer. The position of this instrument was 0,2 under the toe of the pile at final depth, i.e. 18,2 m below groundlevel and 0,5 m besides the pile.

The last mentioned instrument was installed at an angle of 30° with the vertical, in order to place it very near to the pile toe without disturbing the soil close to the pile. The desired level could not be reached; it was placed at a distance of 0,5 m from the pile toe at 18,4 m depth.

The actual measured accelerations had to be processed. The measured signals were rotated in order to get a vertical component, a horizontal component parallel to the direction to the pile and a horizontal component perpendicular to the direction of the pile.

Two soil pressure transducer were placed at a distance of 1 m and 1,8 m from the

Table 4.9. Position of transducers

Transducer	Number	Distance to pile (m)	Depth from surface (m)
Waterpressure	1	0,7	17,1
	2	1,0	17,1
	3	1,0	16,2
	4	0,5	18,4
Accelerometer	1	1,5	17,2
	2	1,5	17,2
	3	0,5	18,4
Soilpressure	1	0,8	18,9
	2	1,6	18,9

centre of the pile at a depth of 19 m below groundlevel. Just before the static loading test.

The complete instrumentation plan is shown in Figure 4.11 and the exact locations are shown in Table 4.9.

Results

During pile driving 21 two-seconds measurements were carried out. Each measurement had contained at least one blow. The sample rate was 3000 Hz. The measurements 10 to 21 were carried out during penetration of the pile toe in the deep sand layer. Figure 4.12 shows the measured acceleration on the moment that the pile toe was at 15 m depth.

Only one water pressure transducer functioned during driving. The pore pressure measured when the pile toe reached a depth of 16,5 m is shown in Figure 4.13.

During the dynamic loading three measurements were carried out. A measurement during the third blow is shown in Figure 4.14.

During the statnamic loading one measurement was carried out. Figure 4.15 shows the measured water pressure. It is noted that the Figures 4.14 and 4.15 show results which were measured above the pile toe level.

A more extensive description of the measurements is published by Hölscher (1995).

4.4.2 *Measurements at the pile shaft*

During the installation of the testpiles 1-5 measurements were performed to check the pile driving parameters:
 – Driving resistance;
 – Maximum compressive stress;
 – Energy transferred to the pile by the hammer.
Therefore the piles were instrumented near the pile top with bolted on acceleration/strain transducers. The measurements were carried out with the TNO FPDS-3 system. The piles were driven with the hydrohammer ICE SC-40. Recorded blow counts are listed in the Tables 4.4a-4.4e.

In the Figures 4.16-4.20 the results of the measurements of driving resistance

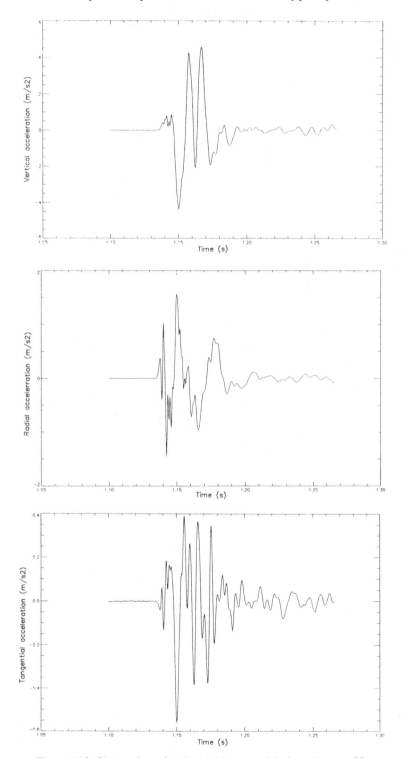

Figure 4.12. Observed accelerations in the ground during a hammer blow.

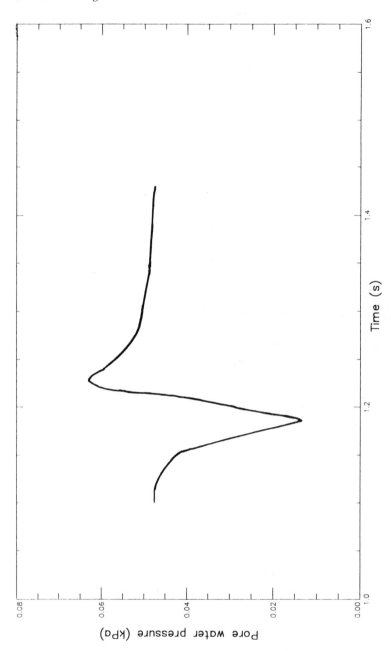

Figure 4.13. Observed waterpressure during a hammer blow.

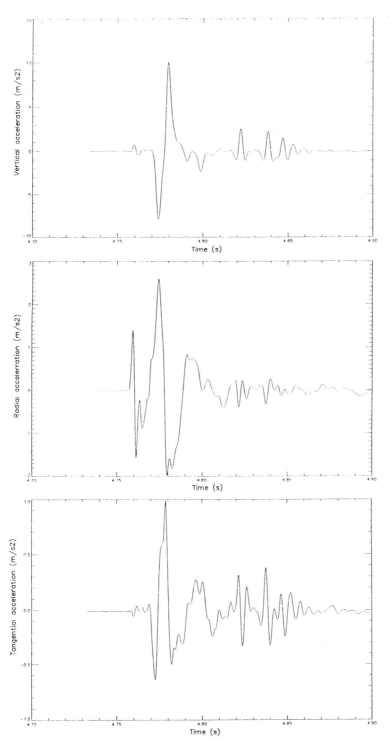

Figure 4.14. Acceleration in the ground during dynamic test loading of pile 3.

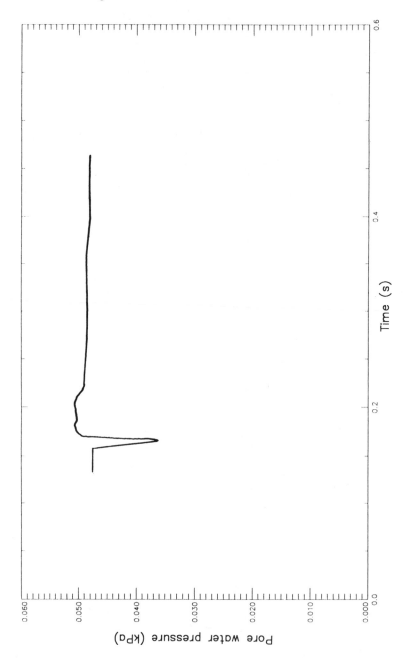

Figure 4.15. Observed porewater pressure in het ground during dynamic load test of pile 3.

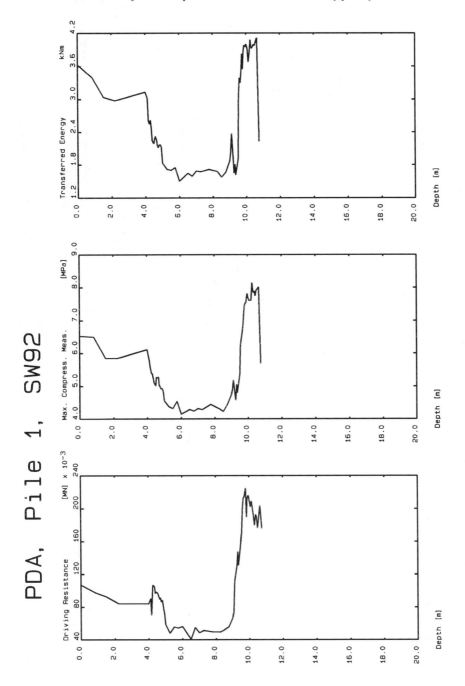

Figure 4.16. Measurements of driving resistance/maximum compressive stress and transferred energy, pile 1.

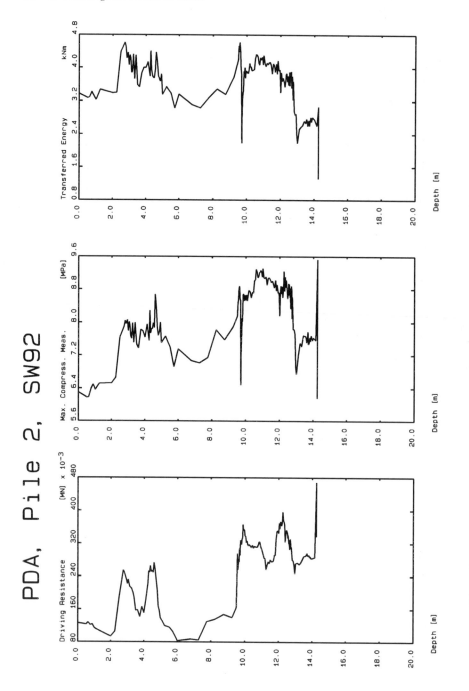

Figure 4.17. Measurements of driving resistance/maximum compressive stress and transferred energy, pile 2.

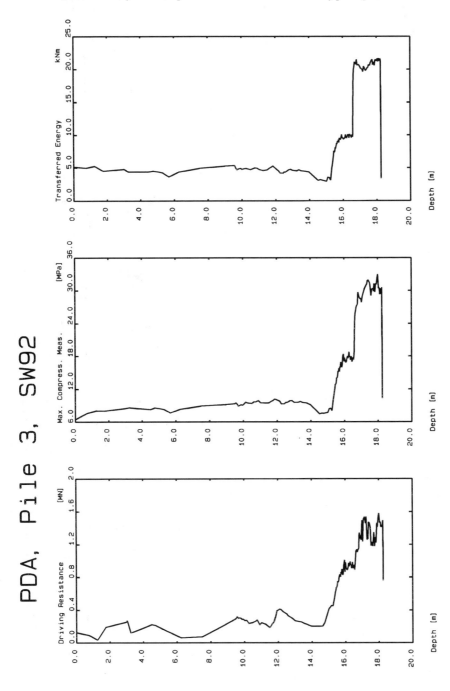

Figure 4.18. Measurements of driving resistance/maximum compressive stress and transferred energy, pile 3.

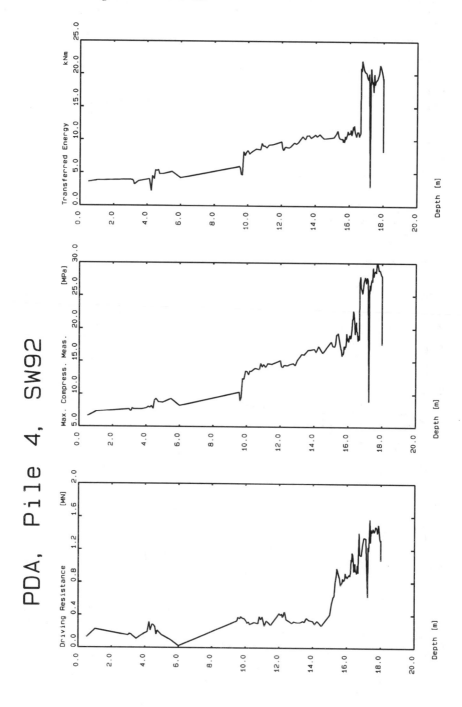

Figure 4.19. Measurements of driving resistance/maximum compressive stress and transferred energy, pile 4.

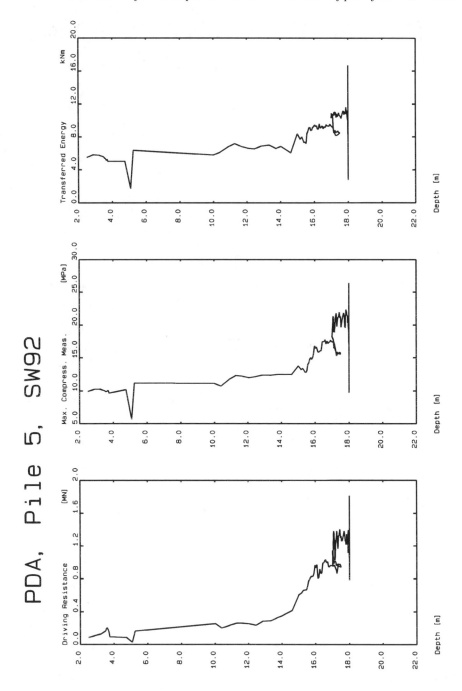

Figure 4.20. Measurements of driving resistance/maximum compressive stress and transferred energy, pile 5.

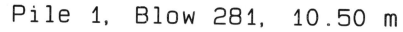

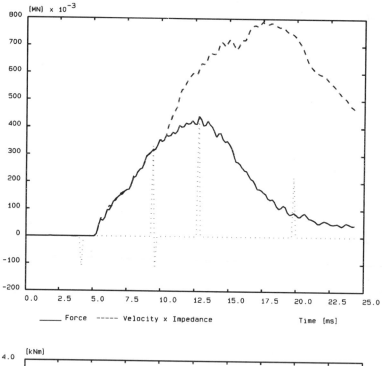

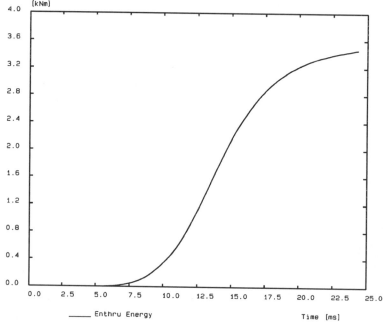

Figure 4.21. Force/velocity and enthru energy under the last blow on pile 1.

Pile 2, Blow 704, 14.25 m

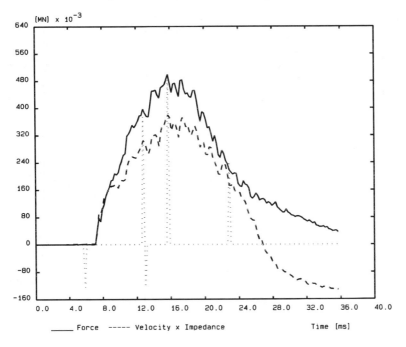

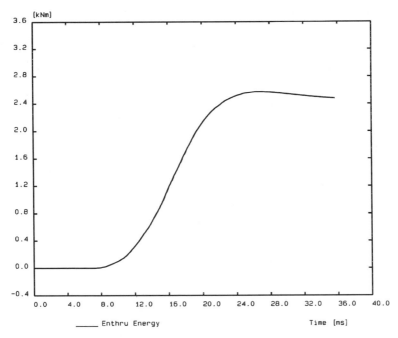

Figure 4.22. Force/velocity and enthru energy under the last blow on pile 2.

Pile 3, Blow 673, 18.25 m

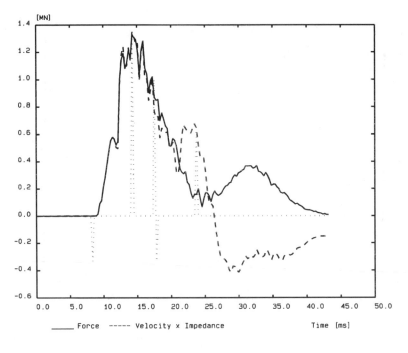

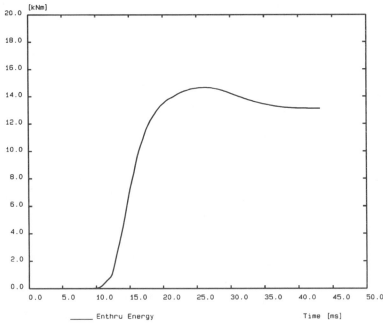

Figure 4.23. Force/velocity and enthru energy under the last blow on pile 3.

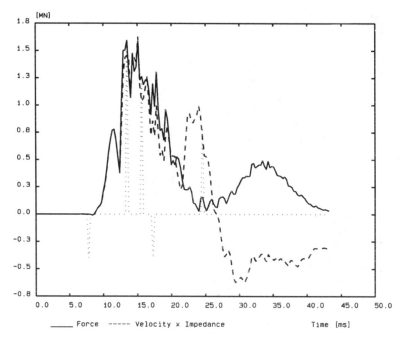

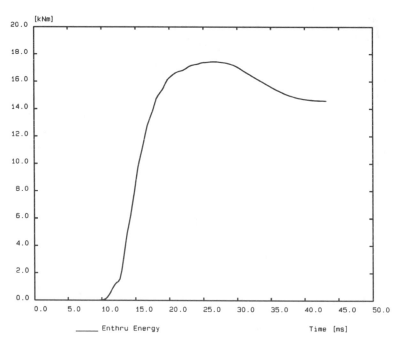

Figure 4.24. Force/velocity and enthru energy under the last blow on pile 4.

Pile 5, Blow 558, 18.00 m

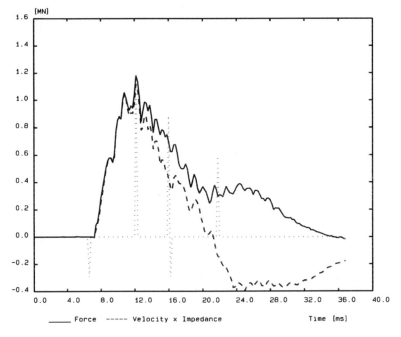

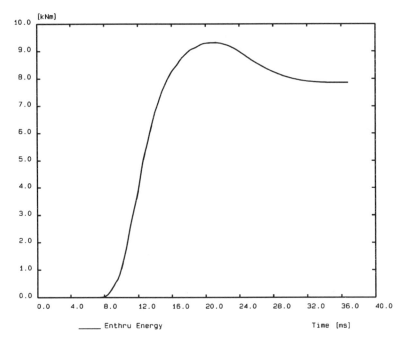

Figure 4.25. Force/velocity and enthru energy under the last blow on pile 5.

maximum compressive stress and transferred energy are shown as a function of depth.

Figures 4.21-4.25 show the impact force, the velocity times impedance records and the enthru energy for the piles 1-5.

4.5 COMPARISON OF THE PREDICTIONS FROM THE DYNAMIC AND KINEMATIC LOAD TESTS WITH THE STATIC LOAD TEST RESULTS

Figures 4.26-4.30 inclusive show the predicted load settlement curves of the piles 1,2,3 and 5 for static loading conditions as derived from the dynamic and the kinetic pile load tests. In each figure the measured static load-settlement curve of the relevant pile is shown too.

Pile 4 was only subjected to a static loading test.

4.6 EVALUATION OF THE DYNAMIC/KINETIC LOAD TEST RESULTS AND CONCLUSIONS

4.6.1 *The contest*

The participants in the contest predicted the load settlement behaviour of the piles 1, 2, 3 and 5 on the basis of the results of the dynamic and kinetic loading of these piles. These predictions were sent to the notary of the Delft University of Technology. He was responsible for keeping the results secret.

The names of the participants were replaced by anonymous labels A-K before the predictions were judged against the measured load-displacement behaviour of the piles 1, 2. 3 and 5.

The best prediction was selected on the basis of this comparison with the static pile load test results taking into consideration the following two objectives:
– Maximum bearing capacity;
– Pile head displacements curve over the range 0-50% of the maximum load of the static load test.

The load which caused a pile head displacement of 10% of the equivalent diameter was considered as the maximum bearing capacity. In this case it means that this criterion concurs with a displacement of 30 mm. In Table 4.11 the maximum bearing capacity of the piles 1-5, as established by this criterion, are summarized.

For the selection of the winners a score between 0-1 was attributed to the predictions for both criteria. The score is based on the extent of the deviation of the predicted value from the measured one. In this way each participant got a total score between 0-2.

4.6.2 *Evaluation of dynamic test results*

Accuracy of the predictions:
The differences, expressed as the ratio between the predicted and measured values

for the maximum bearing capacity and the displacements, are summarized in Tables 4.11 and 4.12.

The maximum bearing capacity is in average overestimated by a factor of approximately 1.2. The variation coefficient is about 0,25.

The displacement of the pilehead at 50% of the maximum load as derived from the static load test, is overestimated by a factor of approximately 1,3. The variation coefficient is about 0,75.

The predicted displacements of the friction pile 2, were relatively inaccurate. This is an unexpected result, because friction piles can usually be analysed rather accurately.

Table 4.10. Maximum bearing capacity following from the displacement criterion of 30 mm.

Pile number	Maximum bearing capacity in kN
1	324
2	595
3	1 117
4	1 210
5	1 215

Table 4.11. Ratio between the predicted and measured values for the maximum bearing capacity.

Participant	Pile 1	Pile 2	Pile 3	Pile 5
A	0,28	0,59	0,70	0,48
B	–	1,14	–	1,0
C	–	1,04	0,98	1,12
D	–	–	–	–
E	–	–	–	1,50
F	1,41	1,19	1,20	1,32
G	1,23	–	–	1,75
H	–	–	–	–
I	1,35	0,95	0,85	–
J	1,60	0,77	0,97	1,19
K	1,32	0,96	0,87	1,19

Table 4.12. Ratio between the predicted and measured displacements at 50% of the ultimate bearing capacity.

Participant	Pile 1	Pile 2	Pile 3	Pile 5
A	–	–	–	–
B	–	0,59	1,32	0,70
C	3,32	0,80	1,53	1,08
D	–	–	–	–
E	–	–	–	1,06
F	1,20	1,10	1,29	1,25
G	2,00	1,06	1,38	1,50
H	0,60	0,63	1,35	1,22
I	1,03	0,79	1,26	0,80
J	1,42	4,46	1,04	0,71
K	1,21	0,74	1,41	1,00

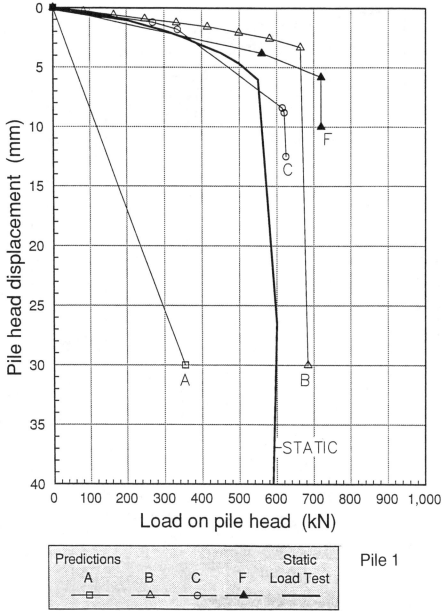

Figure 4.26a. Prediction vs actual behaviour, pile 1.

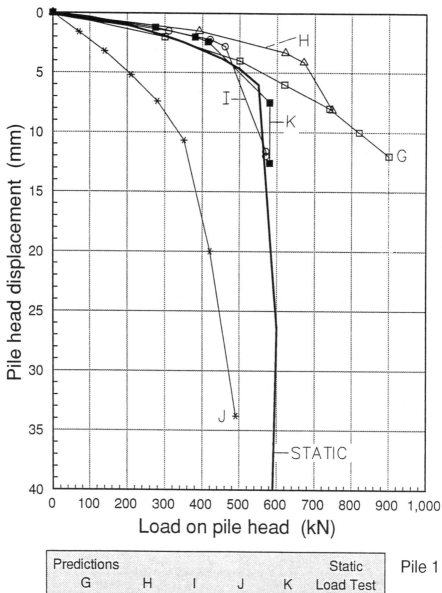

Figure 4.26b. Prediction vs actual behaviour, pile 1.

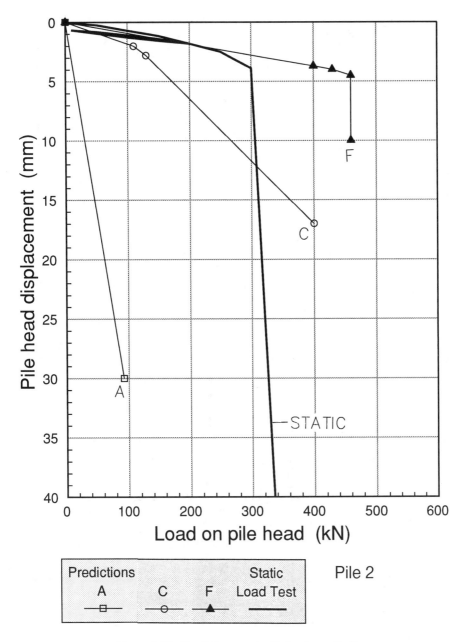

Figure 4.27a. Prediction vs actual behaviour, pile 2.

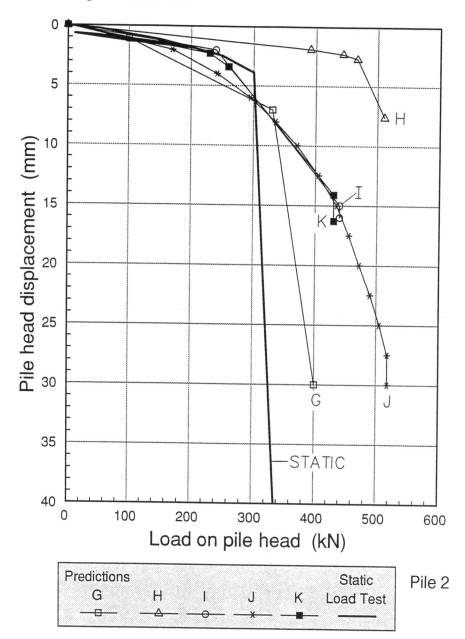

Figure 4.27b. Prediction vs actual behaviour, pile 2.

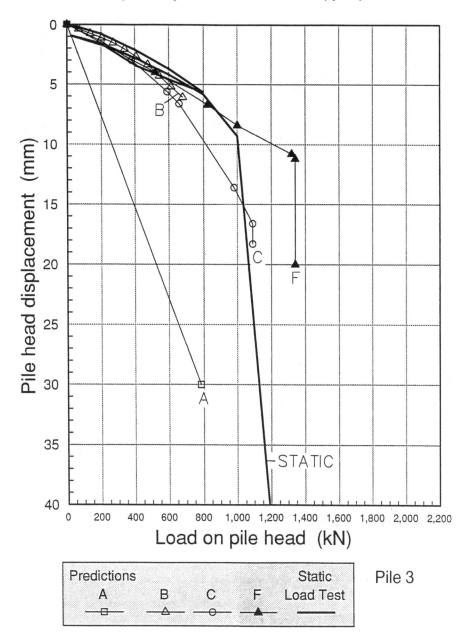

Figure 4.28a. Prediction vs actual behaviour, pile 3.

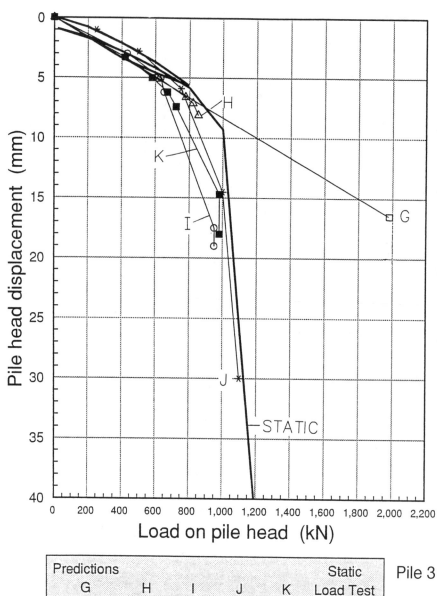

Figure 4.28b. Prediction vs actual behaviour, pile 3.

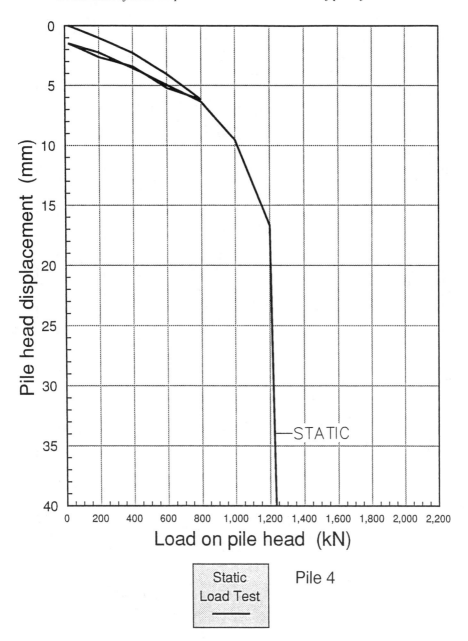

Figure 4.29. Actual behaviour, pile 4.

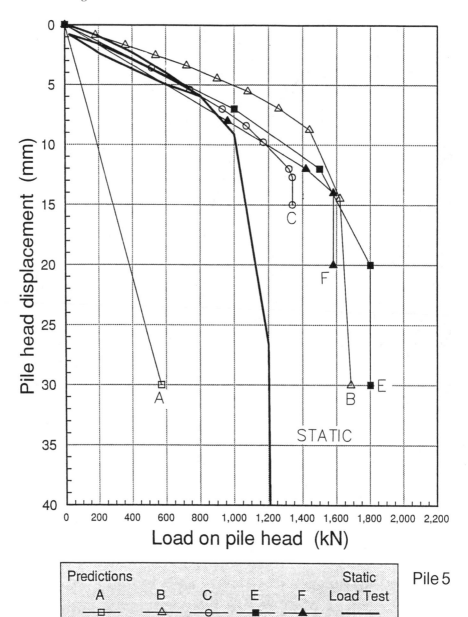

Figure 4.30a. Prediction vs actual behaviour, pile 5.

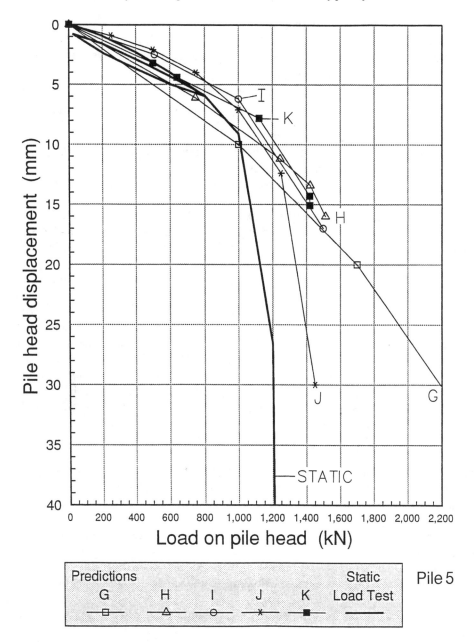

Figure 4.30b. Prediction vs actual behaviour, pile 5.

The predicted maximum bearing capacities for pile 2 were generally good.

The dynamic tests on the similar piles 3 and 5, resulted nevertheless in different predictions of the maximum bearing capacity. All participants overestimated the maximum bearing capacity of pile 5.

For pile 3 the number of overestimations equalled the number of underestimations.

Measurements used by the participants
Most participants used the signals of the dynamic pile load tests for making their predictions for the pile behaviour under static loading conditions.

The kinetic tests results, applied by two participants, are discussed in the next section.

Participant J based his prediction exclusively on CPT-results. Although this analysis method has nothing to do with dynamic testing, the results are included in the Figures 4.15-4.19 for comparison reasons. They are not taken into considerations in the contest.

In Table 4.6 the various measuring systems used by the participants are summarized.

The participants provided only very limited information of the applied measuring system. Many of them did not present the measured signals. The measurements which were available. revealed that only marginal differences between the various measuring systems occurred. Generally they were < 5% for both, velocity and force.

Some participants observed large differences between the force and/or the velocity readings on opposite sides of the pile. These differences must be attributed to eccentricities of the blows. Faulty data acquisition is an unlikely reason because of the good general agreement between the various methods.

Interpretation methods
The predictions of the static pile behaviour was generally made by matching the measured signal of the upward travelling wave with the computation results. The input data provided after matching the data for the prediction of the pile capacity under static loading conditions.

The participants provided only limited information regarding their computer simulations.

Most of the participants have used the lumped mass method. In this method the soil behaviour is modelled by a combination of linear springs, dashpots or friction elements.

One participant used the continuous pile model, combined with lumped soil parameters.

The resemblance between the results of the users of the lumped mass method C, I and K is rather obvious: Although one of these participants used other measured signals, the results show good agreement with the results of the other two, certainly when the differences between the piles are taken into consideration. Consequently it is concluded that the difference between the results is primarily determined by the applied software. The user of the software and the measuring system only have a moderate influence.

Handicaps in connection with the reliability and the accuracy of the predictions
It must be remarked, that the circumstances during the tests were not ideal. But in this respect the test circumstances bear good resemblance with normal practice.

Some of the sources of inaccuracy are discussed below:

– It is obvious that eventual adverse effects of the pile driving on the maximum bearing capacity of the ground in the vicinity of the pile will disappear with time.

The dynamic/kinetic tests took place 3 days after the installation of the testpiles in the ground. In this short period the soil may not have recovered from the driving effects.

The static load tests were executed 2 weeks after the dynamic/kinetic tests. Although the ground may still not yet have completely recovered during this period, it seems reasonable to assume that the recovery already was in a very advanced stage.

Consequently it is obvious to suppose that the maximum bearing capacity of the piles as deduced from the results of the dynamic/kinetic loading tests should be lower than the results found in the static load tests.

Most of the test results show however the contrary (see Table 4.11) In average the maximum bearing capacity, as derived from the dynamic/kinetic loading tests, exceeds the bearing capacity of the piles established by static loading at the displacement criterion of 150 mm, with approximately 10%. The coefficient of variation is about 0,30. The predictions of participant J were not considered in this comparison because he did not base the prediction on the results af dynamic tests.

For the comparison of the displacements at the load criterion of 50% of the maximum bearing capacity of the static load tests, it is found that, in average, the pile head displacements were overestimated by a factor of 1,19. However the variation coefficient is very high and amounts to $v = 0,44$. Again the prediction of participant J is not taken into account.

If only the results of piles 3 and 5, which are placed in the deep and rather strong sand layer, are taken into consideration, the values become:

a) Ratio dynamic/static maximum bearing capacity: 1,11 with coefficient of variation $v = 0,32$;

b) Ratio dynamic/static displacement: 1,21 with coefficient of variation $v = 0,20$.

The differences, as expressed by the rates, are not very conclusive regarding to the influence of the elapsed time between pile driving and testing.

– Participants could not dispose of the pile driving records (blow counts and hammer performance) and the measurements of the strains and accelerations which were taken during pile driving. Normaly these data are, if available, also taken into consideration in the analysis of the results of the dynamic/kinetic tests. But they are not of crucial importance for matching the computation results with the signals measured at the dynamic/kinetic loading test.

– Dynamic/kinetic loading has caused an additional penetration of testpiles 2, 3 and 5 which varied from 10 to 160 mm. Normally this will not have a significant effect on the outcome of the comparison of dynamic/kinetic loading tests with static testing on piles which is confirmed by the comparison of the static load test results of the piles 3 and 5 with the result of the test on pile 4 which was only loaded statically (see Fig. 4.10).

However in the case of pile 1, an aditional penetration of 330 mm under the dynamic/kinetic tests, has been established. This could have had a substantial effect on the conditions of the pile during the static loading test. Consequently this might have hampered the comparison of the dynamic/kinetic tests with the results of the static test to such an extend that no reliable conclusion can be drawn for this pile.

– The differences between the results of the participants may also be attributed to different blows which they used for the analysis. If a number of blows is applied to the pile, the maximum bearing capacity can decrease due to an increase of the pore pressures in the ground. So under the first blow the pile may meet a higher resistance of the ground than after a number of successive blows.

At low energy levels an increased drop height may lead to a higher resistance.

Participant F stated that his interpretation was hampered by the length of the stress wave.

– The result of a static load test also depends to some extent on the loading procedure. The time of keeping the load constant during the various stages of the test will influence the results, in particular the displacements. The procedure of the static tests contained load steps of 1 hour duration. This is short. Creep effects cannot become effective. This may have yielded to a relatively stiff behaviour in comparison with normal practice. However the same applies to the dynamic/kinetic testing procedures.

4.6.3 *Evaluation of the kinetic testing results*

Line E in the graph for pile 5 shows the result of the statnamic test.

The results of the quasi-static tests of participant D, are not given because they were not interpretable.

Details of the interpretation of statnamic tests are presented in various articles in the proceedings of the Stress-wave Conference (Holeyman 1992).

The statnamic test equipment used for the tests, was designed for a maximum bearing capacity ranging from 2 to 8 MN. Therefore it was not very suitable for the tests on piles 1 and 2. Also the small cross sectional area of the piles caused some problems in setting up the equipment and in inflicting a purely centric blow.

Details about the quasi-static tests have been published by Doornbos et al in 1994.

4.7 CONCLUSIONS

The tests executed during the Stress-wave Conference at the test site in Delft have shown that dynamic load tests can provide good information for a fast and reasonably reliable assessment of the maximum bearing capacity and the displacement behaviour of foundation piles. The test procedures, as applied in Delft, reflected more the normal practice than the scientific approach which was in fact the primary goal.

The overall overestimation of the maximum bearing capacity was about 10%; the coefficient of variation 0,3.

The overestimation of the displacement amounted to 20%; coefficient of variation 0,44.

If only the piles 3 and 5, which have a normal pile toe level in the bearing sand stratum, are considered these values become:

– For the maximum bearing capacity, an average overestimaton of 11% with a coefficient of variation $v = 0,32$;

– For the displacement, an overestimation of 21% with a coefficient of variation of $v = 0,20$.

Athough the comparison is hampered by a number of draw backs they obviously were not of decisive importance.

The consistency between the interpretation results of different users of the same software is good.

The interpretation method and the type of software used for the conversion of the results of the dynamic tests into static pile behaviour seems to be more important than differences between measuring systems and users.

It has been demonstrated that local experience is not a decisive factor to get to a reliable prediction of static pile behaviour from dynamic load test results.

The trials have shown that it is worthwhile to stimulate the research and development efforts in the area of kinetic testing.

REFERENCES

Bermingham, P. & Janes, M., 1989. An innovative approach to load testing of high capacity piles. *Proc. Int. Conf. on Piling and Deep Foundations*, London.

Doornbos, S., Revoort, E., Schoo, O. & Tirkkonen, O., 1994. Comparison of Pile Loading Tests and The Phenomenon of Heave at Sachsen Paper Mill Eilenburg, *Proc. 5th International Conference on Piling and Deep Foundations,* 13-15 June, pp 4.2.1-4.2.12.

Delft Geotechnics, 1992. Report CO 335630/15 of August 1992; Factual Report, 4th Int. Stress Wave Conf. – Static pile load tests at Delft, The Netherlands.

Gonin, H., Coelus, G. & Leonard, M., 1984. Theory and performance of a new dynamic method of pile testing. *Proc. 2nd Int. Conf. Application of Stress Wave Theory to Piles*, Stockholm.

Holeyman, A.E., 1992, Keynote Lecture: Technology of Pile Dynamic Testing, *Proc. 4th Conf. Application of Stresswave Theory to Piles*, The Hague

Middendorp. P., Bermingham. P. & Kuiper.B., 1992, Statnamic Load Testing of Foundation Piles, *Proc. 4th Conf. Application of Stresswave Theory to Piles*, The Hague

Hölscher, P., 1995. Dynamical response of saturated and dry soils PhD. thesis, Delft University of Technology, Department of Civil Engineering, Delft University Press, Delft, the Netherlands, February 1995, ISBN 90-407-1073-2

CHAPTER 5

Pile driving prediction contest

F.T.M. GOZELING
Fugro Engineers BV, Leidschendam, Netherlands

E.G. VAN DER VELDE
Fugro Engineers BV, Leidschendam, Netherlands

5.1 INTRODUCTION

Four different hammers drove simultaneously identical precast concrete piles at the test site at Delft University of Technology. This was done during the 4th Stress Wave Conference on September 22, 1992. The hammers were operated by four Dutch piling contractors. This operation formed part of a pile driving prediction contest and hence pile driving was accurately monitored. For this purpose all piles were instrumented with accelerometers and strain gauges. Four identical sets, supplied and operated by TNO Building and Construction Research, provided the monitoring data for each pile of the pile driving prediction contest. Actual monitored data were subsequently compared with predictions made by 11 anonymous participants from all over the world. Two winners were selected on the basis of their interpolated/extrapolated predictions on stress levels, and blowcount predictions closest to the actual values. The winners are ranked as follows:
- Delft Geotechnics, The Netherlands;
- University of British Colombia, Canada.

Piling contractor Oudenallen with the Delmag D-25 diesel hammer was declared as the overall winner.

This report presents and discusses the results of the pile monitoring, the predictions and the comparison of the predictions with the measured data.

5.2 OBJECTIVES

The objectives of the pile driving demonstration on September 22, 1993 were two fold. Firstly, it was a test on the accuracy of pile driving predictions by various methods. To enhance this test, four different hammers were used to drive 20 m long slender prestressed concrete piles simultaneously to 19 m penetration. Blowcounts, pile stresses and 'hammer energy transferred into the pile' (enthru energy) were monitored continuously. By comparing various analysis methods and driving equipment a general impression of the accuracy of most commonly used pile driving prediction methods and the efficiency of the driving equipment could be obtained. Secondly, it was a contest between the four contractors to drive the piles in the best possible way without damaging them. This pile driving contest also served a less professional goal; the selection of the best pile driving prediction made by the participants.

5.3 SOIL CONDITIONS

In Chapter 2 a review is given of the soil conditions of the test site. At the vicinity of each pile location a Dutch Cone Penetration Test (CPT) was performed in advance. On the test site one Begemann boring was made. Table 5.1 gives the average soil profile. In general the soil is quite homogeneous over the area of the test site.

Table 5.1. Average soil stratigraphy.

Layer no.	Depth below NAP (m) from ... to ...	Layer description	Average cone resistance (MPa)
1	0.8/1.0-3.4	Clay, silty	1.0
2	3.4-5.7	Sand, slightly silty, with clay layers	4.3
3	5.7-7.0	Peat	0.5
4	7.0-9.8	Clay, slightly silty, slightly organic	0.5
5	9.8-13.7	Sand, slightly silty	4.2
6	13.7-14.9	Sand, silty, with clay layers	1.5
7	14.9-15.6/15.9	Peat, clayey	1.0
8	15.6/15.9-30+	Sand	11 - 18

The ground water level at the test site was 0.6 m below ground level. Ground level was at 0.5 to 1 m below NAP.

5.4 HAMMER, PILE AND INSTRUMENTATION DATA

5.4.1 *Pile data*

The cross section of the four prefabricated square 20 m long slender prestressed concrete piles was 250 mm * 250 mm. The final pile toe penetration was 19 m below ground level (approx. 20 m –NAP). This depth was selected such that the piles would meet substantial driving resistance. The quality of the concrete was B55. This stands for concrete with a characteristic compressive strength after 28 days of 55 MPa. The piles were fabricated on May 13, 1992. In each pile 6 rods of 9.3 mm diameter (steel quality: FeP 1860) were applied to prestress the concrete section to 4.8 MPa. The specified pile Young's modulus (E) was 36,000 MPa.

5.4.2 *Driving equipment*

Each of the four pile driving rigs was equipped with a different type of hammer. These were a single acting Delmag D-25 diesel hammer, a double acting ICE D-640 diesel hammer, a Menck MHF 3-3 hydraulic hammer and an IHC SC-40 hydraulic

Table 5.2. Assigned hammer locations and operators.

Location	Hammer	Pile driving rig	Contractor
04	Delmag D-25	Kobelco 7045 XLR	Oudenallen
05	Menck MHF 3-3	Linkbelt LS120	Guis
06	ICE D-640	Hitachi KH150-3HD	Hoogwerff
07	IHC SC-40	Hitachi 150	Vroom

Table 5.3. Hammer specifications.

	Delmag D-25	Menck MHF 3-3	ICE D-640	IHC SC-40
Energy range (kNm)	40-79	2-30	35-55	4-38
Total ram mass (kg)	2500	3200	2730	2550
Blow rate (bl/min)	37-52	50	74-77	50
Hammer mass (kg)	5610	5500	7070	5100

hammer. In Table 5.2 the rigs and hammer combinations which were used at the test locations are summarised.

All hammers except the IHC SC-40 have the same helmet configuration with similar cap blocks and cushions.

The hammer operating data and pile cap configurations are given in Figure 5.1. The hammer specifications are presented in Table 5.3.

5.4.3 *Instrumentation equipment*

The instrumentation of the piles is shown in Figure 5.2.

Each of the four piles was instrumented with two accelerometers and two strain gauges in opposite pairs at 0.5 m below pile head level. These sensors are of a dual type as designed by TNO Building and Construction Research; one strain gauge and one accelerometer are combined in one housing which was secured with anchor bolts to the pile. The accelerometer is of the piezo-resistive type. The sensors were connected through an umbilical cable to the data acquisition equipment. The data were acquired by the Foundation Pile Diagnostic System-3. The acceleration signals were converted to velocity signals; the strain signals to force signals. With these signals, wave analysis was performed. Hence, maximum compression stresses, energy in the pile, eccentricity of the impact, and more, can be determined for each blow.

In addition to these routine pile head measurements, two strain gauges were glued to opposite faces of the pile shaft at a level of 9 m below pile head. In this way tensile stresses at this level were measured during driving. Each strain gauge was a quarter of a Wheatstone bridge. The signals were amplified and subsequently recorded on analogue tape for future analysis. They were processed by the Foundation Pile Diagnostic System-3. The measured strain signals were multiplied by the pile Young's Modulus to obtain stresses. The average of the maximum tensile stress from the two opposite sides of the pile shaft at 9 m below the pile head has been determined for each hammer blow.

During driving of the piles, vibration measurements were also carried out at the ground surface in the surroundings (see Chapter 7).

		Delmag D-25 single acting diesel	Menck MHF 3-3	ICE D-640	IHC SC-40
Energy: max.	(kJ)	79	30	55	38
min.	(kJ)	40	2	35	4
Ram mass total	(kg)	2500	3200	2730	2550
Impact block: mass	(kg)	760	n.a.	660	n.a.
length	(m)	0,705	n.a.	0,515	n.a.
cross area	(m^2)	0,138	n.a.	0,163	n.a.
Peak force of explosion	(kN)	1304	n.a.	1400	n.a.
Cap arrangement		A	A	A	B

Remarks:
a. Energy of diesel hammer is rated energy as mentioned in brochure of manufacturer.
b. Energy of hydraulic hammers is net energy, delivered to the anvil, as mentioned in brochure of manufacturer.

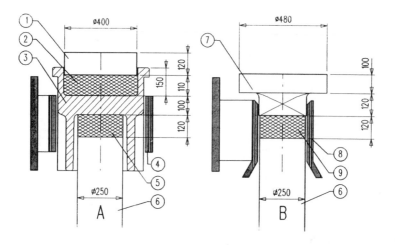

No.	Parts	Material	Weights Kg	Cushion stiffness kN/mm
1	Anvil	Steel	120	-
2	Cushion	Hardwood	-	1650
3	Cap	Steel	280	-
4	Guidepart	Steel	-	-
5	Pile cushion	Softwood	-	110
6	Pile	Concrete	-	-
7	Anvil	Steel	200	-
8	Guidepart	Steel	-	-
9	Pile cushion	Softwood	-	110

Figure 5.1. Hammer and pile cap information for drivability predictions.

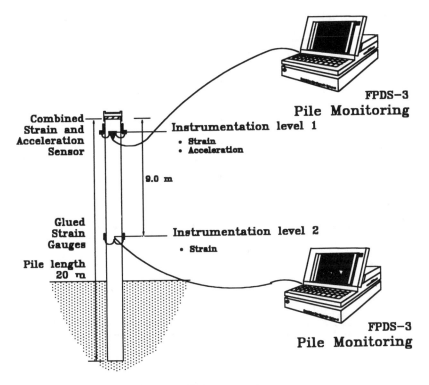

Figure 5.2. Pile intrumentation set-up.

5.5 PILE DRIVING OBSERVATIONS

The piles were marked each 0.25 m to log the blowcounts and to label the monitoring data with penetration. The piles were installed on September 22, 1992, the Demonstration Day of the 4th International Stress Wave Conference. The pile driving was observed by four field engineers who recorded the blowcount per 0.25 m penetration of the piles. The passing of each driving interval was relayed to the TNO monitoring crew. The pile penetrations due to the weight of the pile and hammer placement ranged from 2.5 m for the hydraulic hammers, IHC SC-40 and Menck MHF 3-3, to 5 and 9.5 m for the diesel hammers ICE D-640 and Delmag D-25 respectively. The driving of the piles was started exactly at the same time. The objective was to reach the 19 m target penetration as fast as possible without damaging the piles.

Plots of blowcounts for all four hammers are given in Figure 5.3. Generally low blowcounts (1 to 5 blows per 0.25 m) to a pile penetration of 15.5 m (NAP – 16.4 m), were observed, accompanied by rapid penetrations through the peat and clay layers. The initial lower energy setting of the Menck MHF 3-3 at shallow penetrations resulted in some peak blowcount values of 8 to 14 blows per 0.25 m. The blowcount curves for the IHC SC-40 and Menck MHF 3-3 show, apart from a slight penetration shift, similar trends. Below a penetration of 15.5 m the blowcounts for each pile increased sharply to a minimum of 12 blows per 0.25 m and a maximum of 26 blows per 0.25 cm for the Delmag D-25 and Menck MHF 3-3 respectively. The blowcount

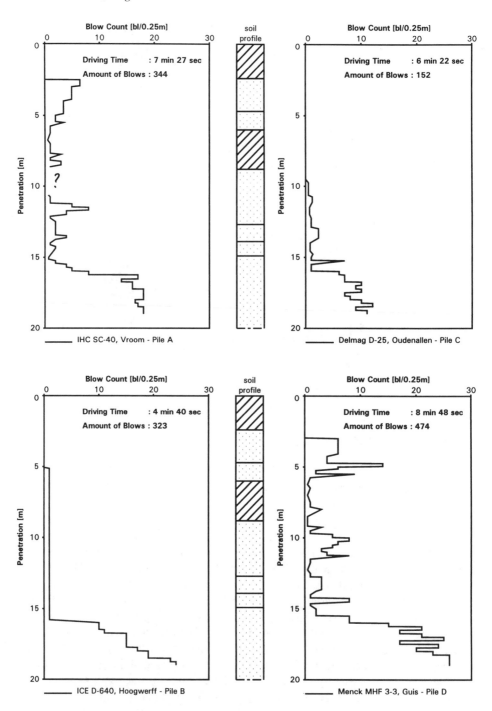

Figure 5.3. Blow-count versus pile penetration.

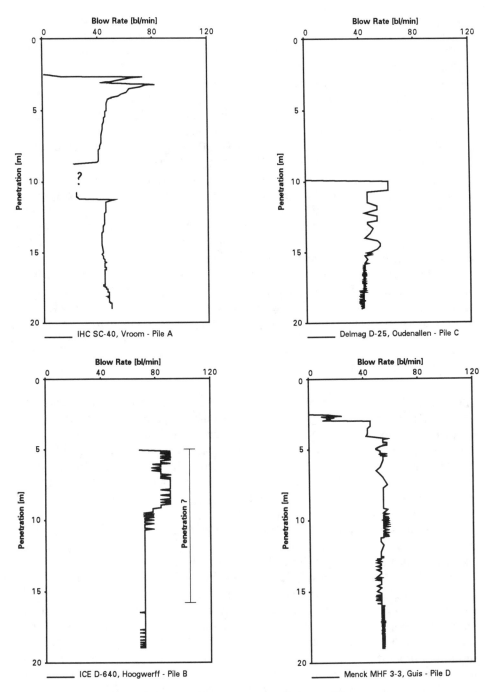

Figure 5.4. Blow rate versus pile penetration.

readings were unreliable for the ICE D-640 diesel hammer to 15.5 m penetration, and for the IHC SC-40 hydraulic hammer between 8.5 and 10 m penetration.

The observed pile driving behaviour generally agrees with what was expected on the basis of the soil stratigraphy (see Chapter 2). Nevertheless, derivation of soil resistance during driving from observed blowcounts should be done with great caution.

Figure 5.4 shows the blow rate profiles (number of blows per minute). The blow rates were mainly operator dependent, showing generally high values during initial driving. However, after some meters of driving the energy/pump settings were increased and all hammers, except for the Menck MHF 3-3, showed a decrease in blow rate. Blow rates decreased within a range of 15 to 30%. The most constant blow rate profile was for the Menck MHF 3-3 hydraulic hammer with an average blow rate of 53 blows per minute. Considerable blow rate fluctuations were observed between 10 and 15 m penetration for the Delmag D-25 diesel hammer as a result of starting problems when no substantial soil resistance was encountered. The blowcount of the Delmag D-25 diesel hammer for pile penetrations in the deeper sand layer were substantially lower than the blowcounts of the other three hammers.

The target depth was reached first by the ICE D-640 hammer. However, Oudenallen with the Delmag D-25 was declared as the overall winner after verification of many other aspects. The total driving time and total number of blows for each hammer are added to the blowcount versus penetration graphs on Figure 5.3.

5.6 INSTRUMENTATION RESULTS

Pile driving operations were monitored over the entire depth driven to quantify the pile driving process. The results of the instrumentation programme are presented in Figures 5.5-5.9 inclusive. They show the following data versus pile penetration:
  – Enthru energy (hammer energy transferred into the pile) in kNm (Fig. 5.5);
  – Maximum average compr. stress at 0.5 m below pile head in MPa (Fig. 5.6);
  – Maximum average tensile stress at 9.0 m below pile head in Mpa (Fig. 5.7);
  – Maximum impact force in MN (Fig. 5.8);
  – Dynamic soil resistance in MN (Fig. 5.9).

The tensile stress measurements, halfway down the pile, were limited up to the depth when the strain gauges penetrated into the soil and the cable was pulled off. This occurred at depths ranging from 10.5 to 16 m (Fig. 5.7).

The instrumentation results for each hammer are briefly discussed below.

The Delmag D-25 diesel hammer experienced start-up problems between the self weight penetration of 9.5 and 15 m penetration. This resulted in fluctuations from 30 up to 50% from the mean values in the energy and stress measurements. At each restart a peak value was recorded during the first blow, followed by lower values as a result of a drop in hammer performance. The increase in enthru energy and stresses became evident with increasing soil resistance below 15 m penetration. Above phenomena are in line with general experience with single acting diesel hammers which work more efficiently when substantial soil resistances are met.

The ICE D-640 double acting diesel hammer provided an essentially constant low energy level during initial driving. A rapid penetration increase was observed between 5 and 16 m depth. Penetration readings over this depth range were therefore

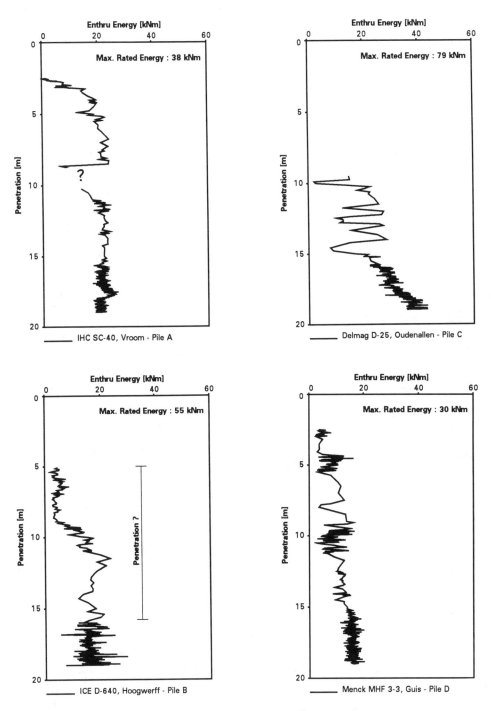

Figure 5.5. Enthru energy versus pile penetration.

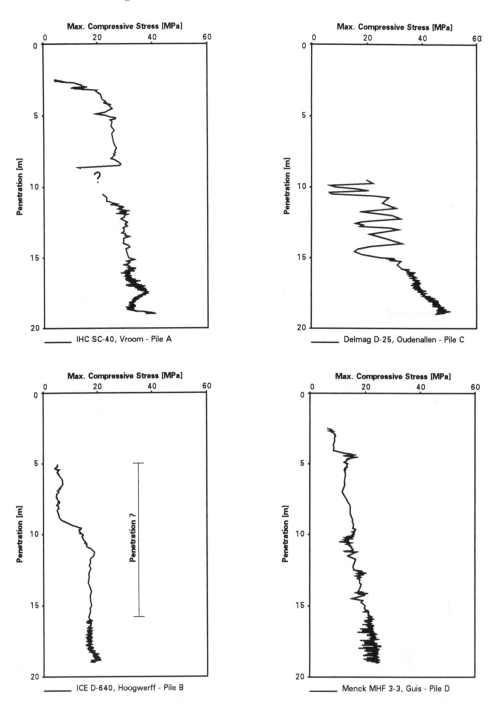

Figure 5.6. Maximum compressive stress versus pile penetration.

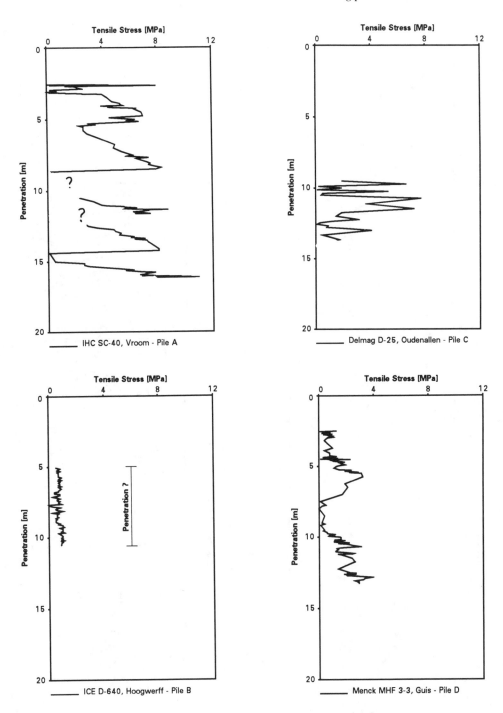

Figure 5.7. Maximum tensile stress versus pile penetration.

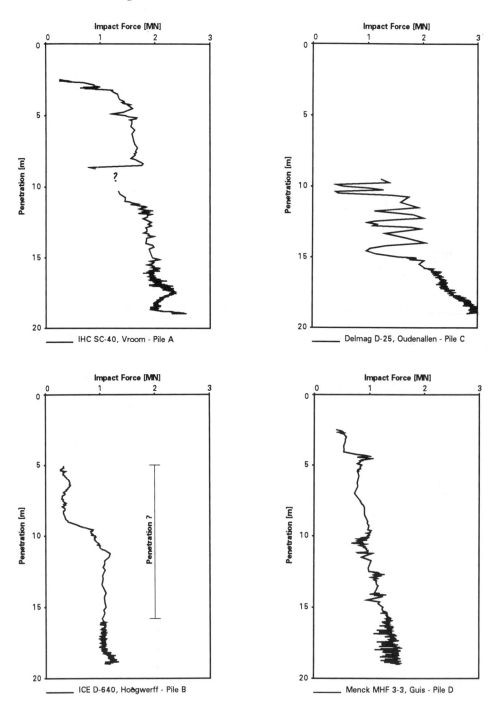

Figure 5.8. Maximum impact force versus pile penetration.

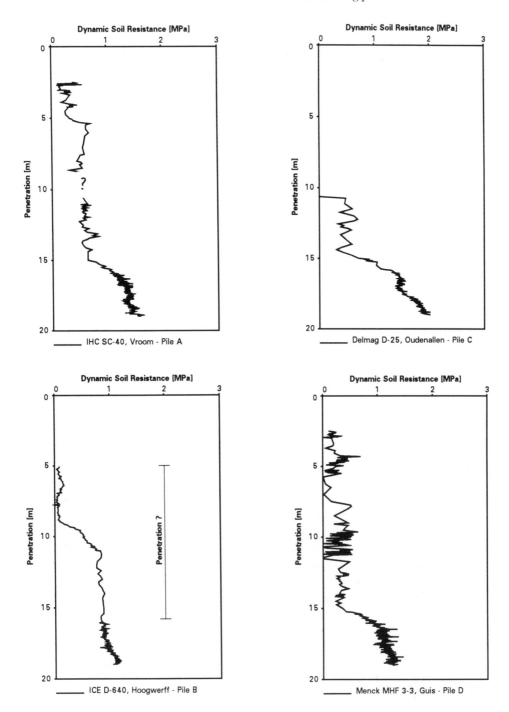

Figure 5.9. Dynamic soil resistance versus pile penetration.

Table 5.4. Pile instrumentation results.

Pile hammer	A IHC SC-40	B ICE D-640	C Delmag D-25	D Menck MHF3-3
Penetration interval (m) – GL	Average Enthru Energy (kNm)			
7.75-8.25	22	4	–	4
9.75-10.25	15	12	10	9
18.50-19.00	21	17	39	16
Penetration interval (m) – GL	Average Blowcount (Blows/0.25 m)			
7.75-8.25	2	2	-	2
9.75-10.25	1	2	1	10
18.50-19.00	18	22	10	26
Penetration interval (m) – GL	Average of max. compressive pile stress at 0.5 m below the pile head (MPa)			
7.75-8.25	27	6	–	14
9.75-10.25	22	16	21	15
18.50-19.00	34	19	47	23
Penetration interval (m) – GL	Average of max. tensile pile stress at 9 m below the pile head (MPa)*			
7.75-8.25	8	1	n.a	0
9.75-10.25	2	1	4	1
18.50-19.00	n.a	n.a	n.a	n.a
Penetration interval (m) – GL	Dynamic soil resistance (MN)			
7.75-8.25	0.5	0.1	n.a	0.5
9.75-10.25	0.5	0.5	n.a	0.3
18.50-19.00	1.5	1.1	1.9	1.3

*Unreliable data.

not accurate. Enthru energy values were mainly operator dependent and hardly influenced by soil resistance. The blowcount increase in the bottom sand layer was more gradual than for the other hammers. Compressive stresses and direct related impact forces remained fairly low and constant between 11 and 18 m penetration. The final meter of driving showed some increase in compressive stress. The observed tensile stresses were low (≤ 1 MPa).

The IHC SC-40 hammer enthru energy remained more or less constant with depth. The compressive stress and impact force gradually increased with depth. This may have been due to the hardening of the cushion. The final meter, however, showed a drop in aforementioned driving characteristics, due to a lower energy setting. Tensile stress measurements showed large variations to relatively deep pile penetrations (strain gauges in soil). These measurements are considered to be unreliable in comparison with the other observations.

The Menck MHF 3-3 hydraulic hammer energy and stress levels generally increased with depth. Somewhat more scatter in data were observed during driving with higher blowcounts in the sand strata. Maximum tensile stresses appear to be closely related with blowcount; similar trends in both measurements can be observed.

It can be assumed that soil conditions and piles at the four locations are comparable. Considering the differences in dynamic resistances (Fig. 5.9), it may be concluded that the dynamic resistance depends on the hammer or that the applied model is not correct. The soil response depends on the type of blow (force-time diagram) and the frequency. The calculation model does not simulate the full complexity of the response. Interpretations of dynamic resistances should therefore be considered with caution.

The dynamic soil resistance has been calculated according to the conventional impedance technique as a summation of dynamic shaft resistance and dynamic toe resistance. Details of the analysis method are given by Beringen et al. (1980) and Rausche et al. (1971).

Typical force-time histories of blows for all hammers were provided by TNO. They are included in the Appendix. The pile head strain gauges and accelerometers generally provided satisfactory signals. The acceleration measurements provide relatively low force data for the diesel hammers. Force-time histories are given for three penetration levels, i.e. 10/12 m, 15 m and 18.75 m – ground level. The energy has been calculated by integrating in time the force-strain signal multiplied by the velocity signal (Beringen et al. 1980).

The key monitoring results which are of particular interest for the pile driving prediction contest are tabulated in Table 5.4. All values have been averaged over the given penetration intervals.

5.7 PILE DRIVING PREDICTION CONTEST

5.7.1 *Participants*

Invitations to participate in the pile driving prediction contest were submitted worldwide early 1992. Interested parties were subsequently provided with an information package containing detailed hammer, pile and soil data. Eleven predictions were received before May 15, 1992; 6 from The Netherlands, 2 from the United States of America and 1 each from Brazil, Canada and Finland. The participants (in random order) were:
- University of British Colombia, Canada;
- Delft Geotechnics, The Netherlands;
- Fugro-McClelland Engineers BV, The Netherlands;
- Goble, Rausche, Likins and Associates, Inc., Mr. M. Svinkin, USA;
- IPT Foundation Consultants, Finland;
- Nederhorst Grondtechniek, The Netherlands;
- Ballast Nedam BV, The Netherlands;
- TNO-Building and Construction Research, The Netherlands;
- Prof. A.F. van Weele, The Netherlands;
- Goble, Rausche, Likins and Associates, Inc., Mr. S.K. Abe, USA;
- Petrobras, Brazil.

The predictions of participants labelled 1 to 11 were kept by the organizing committee. Anonymously labelled copies were made available to Fugro Engineers BV, who compared the predictions with the pile instrumentation results.

5.7.2 *Prediction procedures and required predictions*

Participants were asked to provide the following predictions for each hammer-pile combination:

1. Continuous record of predicted number of blows per 0.25 m pile tip penetration;

2. The average of the maximum compressive stress in the pile at 0.5 m below the pile head for the following 0.5 m long penetration intervals:
   – 7.75-8.25 m,
   – 9.75-10.25 m,
   – 18.50-19 m;

3. The average of the maximum tensile stress in the pile at 9 m below the pile head for the penetration intervals:
   – 7.75-8.25 m,
   – 9.75-10.25 m.

Both the maximum compressive and tensile stresses refer to the stresses measured after mounting the transducers. This implies that the prestress in the pile shaft of 4.8 MPa is excluded.

The above predictions were requested for each hammer operating at assumed 'transferred energy' (enthru energy) levels of 5 kNm, 12.5 kNm and 20 kNm. Therefore the predictions had to be interpolated or extrapolated to actual energy levels determined from the pile instrumentation data for comparison with actually observed data. Subsequently the best predictions for each hammer and the overall best prediction were determined. Subsequently the names of the participants who made these predictions were revealed.

5.7.3 *Prediction results*

Nine participants gave predictions for all four hammers and two participants submitted predictions for two hammers only. Several participants commented that the diesel hammers would not operate at the lowest specified energy of 5 kNm. Unfortunately, descriptions of the analysis method were often vague or not presented. The majority of the participants appear to have assessed pile friction and end bearing from direct empirical correlations with cone resistance. The driving resistance calculated as the summation of friction and toe resistance is a static soil resistance. Dynamic soil resistance during driving are calculated in wave equation programs by incorporating soil damping. Various versions of the WEAP computer program (Goble et al. 1976) and (Smith 1960) were used by most participants to derive blowcounts and pile stresses from soil resistance, pile and hammer data. One participant stated that no calculations had been made, but had based his 2 sets of predictions on experience only.

Table 5.5 summarizes the available information on hammer and soil modelling, and wave equation models and programs used by the eleven participants.

It was decided before the contest to limit the amount of penetration intervals and energy levels to be used. A full comparison for all four hammers was not feasible

Table 5.5. Analysis models.

Participant	Hammer model	Soil model	Wave equation Model/program
1	WEAP 1992	Koppejan	WEAP 1992
2	WEAP 1986/inhouse data	Koppejan/q_c based friction	Sprenger-Potma/WEAP 1986
3	WEAP	–	WEAP
4	WEAP	Nordlund/Tomlinson	WEAP
5	WEAP 1991/inhouse data	Koppejan/q_c based friction	WEAP 1991
6	–	–	–
7	–	–	–
8	–	–	–
9	WEAP 1992	Bustamente/Gianeselli 1982	WEAP 1992
10	–	–	–
11	–	–	–

due to the deep penetration under the self weight for the Delmag D-25, the partial omission of data from the IHC SC-40 and the ICE D-640, and incompleteness of submitted prediction data. The following cases were therefore investigated.

Table 5.6. Investigated predictions.

Penetration interval (m)	Driving parameter	Hammer
7.75-8.25 m	Blowcount/max. tensile stress	IHC SC-40/Menck MHF 3-3
9.75-10.25 m	Blowcount/max. tensile stress	Delmag D-25/Menck MHF 3-3
18.5-19 m	Blowcount/max. compressive stress	IHC SC-40/ICE D-640/Delmag D-25/ Menck MHF 3-3

The predictions of the average blowcount and average maximum compressive stress for an enthru energy level of 20 kNm for all hammers are given in Figures 5.10-5.13 (penetration range 18.5 to 19 m).

The predictions have been converted to the actual observed energy levels for the four hammers for the penetration interval 18.5 to 19 m (Figs 5.15-5.18).

The maximum compressive stress versus enthru energy as measured between 18.5 and 19 m is shown in Figure 5.14. These data reveal little or no correlation between the enthru energy and the maximum compressive stress. This applies especially for the two diesel hammers, where the maximum compressive stress remain constant over a wide range of energy levels.

The observed average blowcounts and maximum compressive stresses between 18.5 and 19 m are shown in Figures 5.15-5.18.

The limited amount of records on tensile stresses does not give a correlation with the enthru energy (Fig. 5.19). The records from the IHC SC-40 are not included; the tensile stresses are most likely not reliable. The best record came from the pile driven with the Menck MHF 3-3. The maximum tensile stress is about 1 to 2 MPa over a wide range of energy levels (3 to 16 kNm).

Predictions of average blowcounts and maximum tensile stresses for penetrations between 7.75 and 10.25 m have been converted to typical actual energy levels for three hammers. These predictions are given in Figures 5.20-5.23.

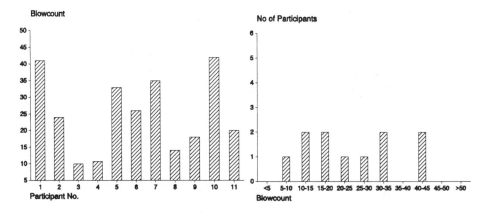

Predicted Average Blowcount, blows/0.25m

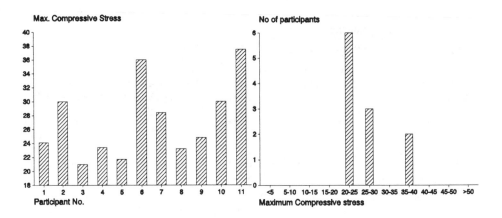

Pred. Avg. Max. Compressive Stress, MPa

Figure 5.10. Predicted average values. Hammer: IHC SC-40; Enthru energy: 20 kNm; Penetration interval: 18.5-19 m.

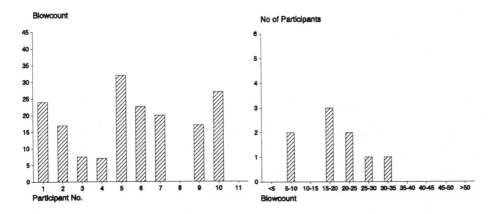

Predicted Average Blowcount, blows/0.25m

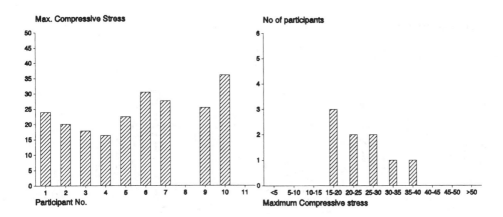

Pred. Avg. Max. Compressive Stress, MPa

Figure 5.11. Predicted average values. Hammer: ICE D-640; Enthru energy: 20 kNm; Penetration interval: 18.5-19 m.

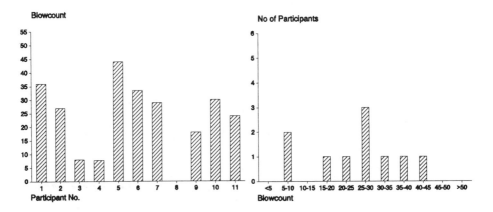

Predicted Average Blowcount, blows/0.25m

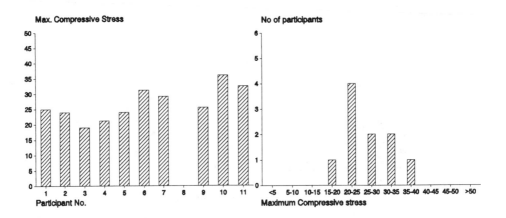

Pred. Avg. Max. Compressive Stress, MPa

Figure 5.12. Predicted average values. Hammer: Delmag D-25; Enthru energy: 20 kNm; Penetration interval: 18.5-19 m.

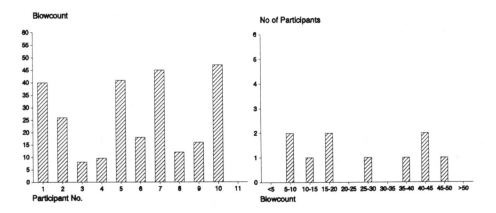

Predicted Average Blowcount, blows/0.25m

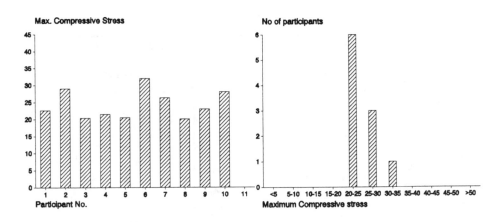

Pred. Avg. Max. Compressive Stress, MPa

Figure 5.13. Predicted average values. Hammer: Menck MHF 3-3; Enthru energy: 20 kNm; Penetration interval: 18.5-19 m.

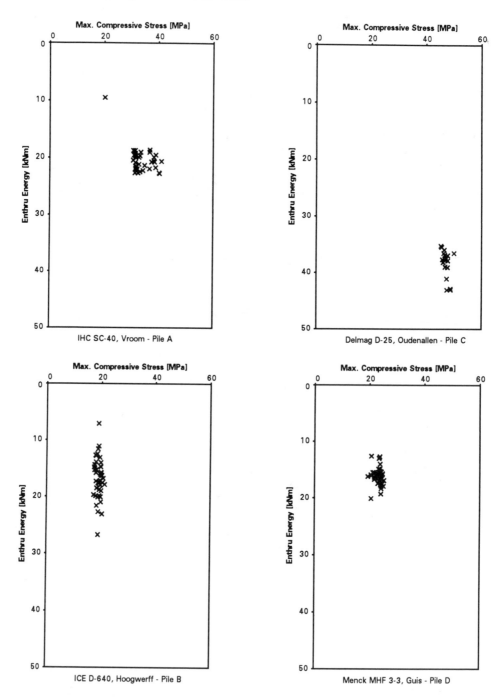

Figure 5.14. Maximum compressive stress versus enthru energy. Penetration interval C: 18.5-19 m.

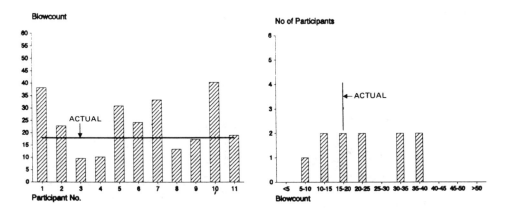

Predicted Average Blowcount, blows/0.25m

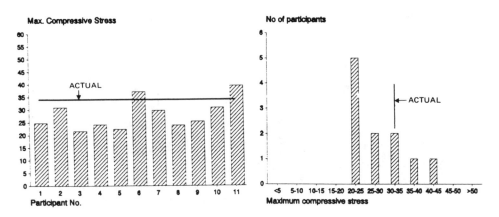

Pred. Avg. Max. Compressive Stress, MPa

Figure 5.15. Extrapolated predictions and actual values. Hammer: IHC SC-40; Enthru energy: 21 kNm; Penetration interval: 18.5-19 m.

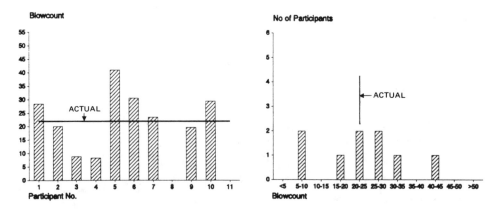

Predicted Average Blowcount, blows/0.25m

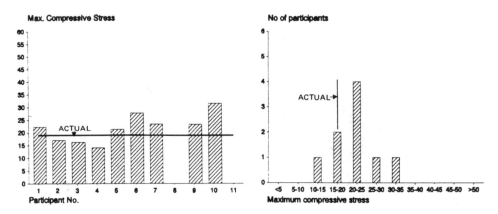

Pred. Avg. Max. Compressive Stress, MPa

Figure 5.16. Interpolated predictions and actual values. Hammer: ICE D-640; Enthru energy: 17 kNm; Penetration interval: 18.5-19 m.

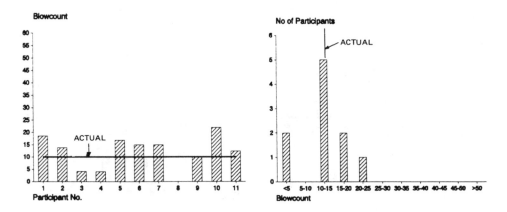

Predicted Average Blowcount, blows/0.25m

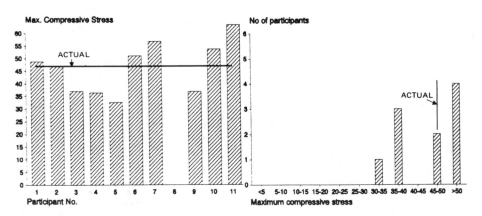

Pred. Avg. Max. Compressive Stress, MPa

Figure 5.17. Extrapolated predictions and actual values. Hammer: Delmag D-25; Enthru energy: 39 kNm; Penetration interval: 18.5-19 m.

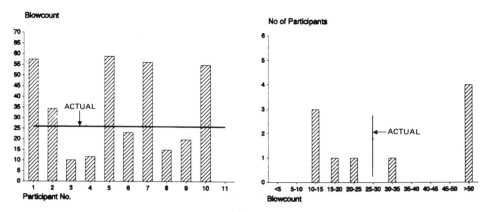

Predicted Average Blowcount, blows/0.25m

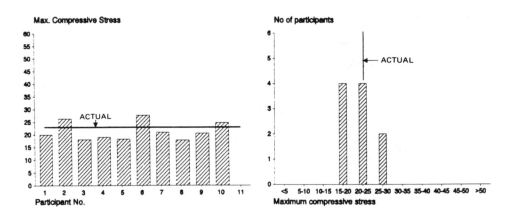

Pred. Avg. Max. Compressive Stress, MPa

Figure 5.18. Interpolated predictions and actual values. Hammer: Menck MHF 3-3; Enthru energy: 16 kNm; Penetration interval: 18.5-19 m.

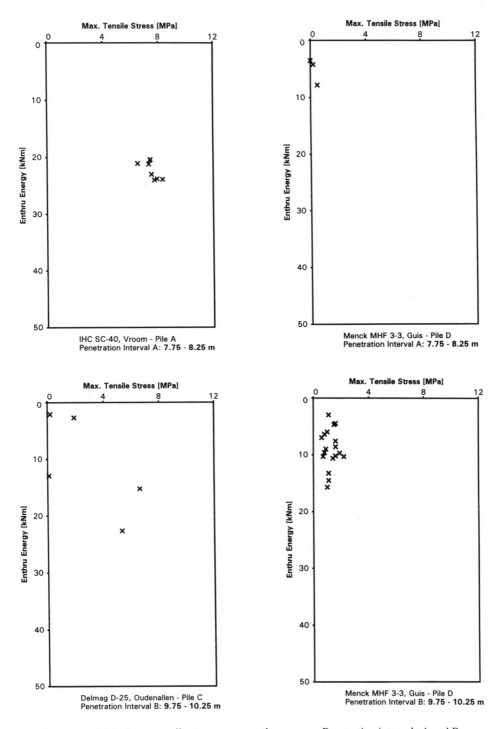

Figure 5.19. Maximum tensile stress versus enthru energy. Penetration intervals A and B.

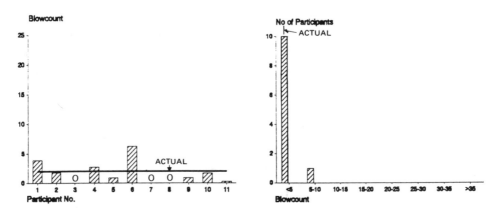

Predicted Average Blowcount, blows/0.25m

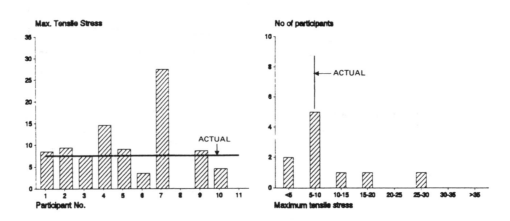

Pred. Avg. Max. Tensile Stress, MPa

Figure 5.20. Extrapolated predictions and actual values. Hammer: IHC SC-40; Enthru energy: 22 kNm; Penetration interval: 7.75-8.25 m.

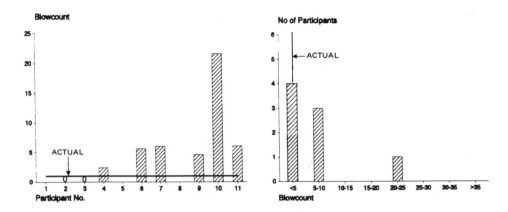

Predicted Average Blowcount, blows/0.25m

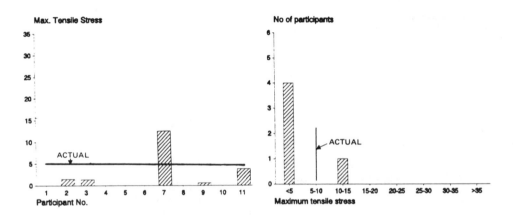

Pred. Avg. Max. Tensile Stress, MPa

Figure 5.21. Interpolated predictions and actual values. Hammer: Delmag D-25; Enthru energy: 15 kNm; Penetration interval: 9.75-10.25 m.

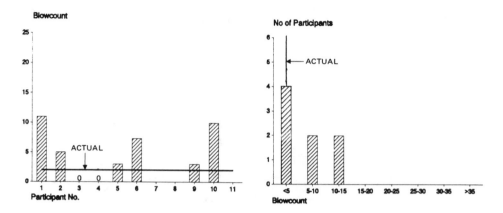

Predicted Average Blowcount, blows/0.25m

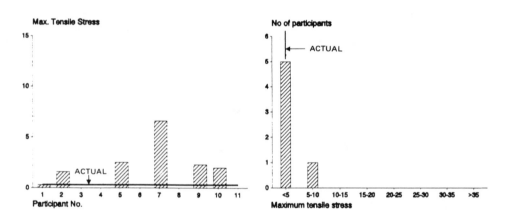

Pred. Avg. Max. Tensile Stress, MPa

Figure 5.22. Interpolated predictions and actual values. Hammer: Menck MHF 3-3; Enthru energy: 5 kNm; Penetration interval: 7.75-8.25 m.

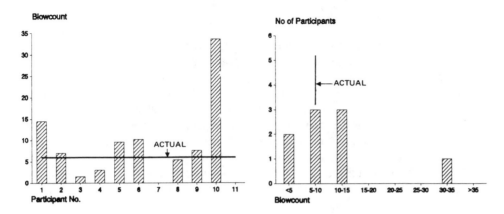

Predicted Average Blowcount, blows/0.25m

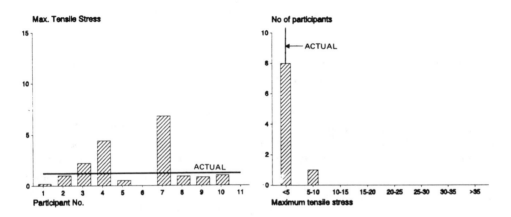

Pred. Avg. Max. Tensile Stress, MPa

Figure 5.23. Interpolated predictions and actual values. Hammer: Menck MHF 3-3; Enthru energy: 9 kNm; Penetration interval: 9.75-10.25 m.

5.8 PREDICTIONS VERSUS OBSERVATIONS

The observations or actual measurements are represented by lines in the bar diagrams of predicted values in Figures 5.15-5.18 and Figures 5.20-5.23.

The predicted blowcounts of the participants reveal large variations. The average of the predictions corresponds reasonably with the observation between 18.5 and 19 m. The blowcounts at shallow depth are generally overestimated.

The data suggest that the predictions per participant are consistently too high or to low for the four hammers. This indicates that the deviations are primarily due to a systematical overestimate or an underestimate of the driving resistance.

Figures 5.15-5.18 indicate clearly that the compressive stress comparisons are underestimated for the hydraulic hammers by the majority of the participants. A scatter both ways was observed for the other hammers.

No definite trend can be concluded from the tensile stress predictions and observations shown in Figures 5.20-5.23. The limited amount of reliable data indicate that the maximum tensile stress in the pile is independent of the energy level.

5.9 GENERAL CONCLUSIONS

The main conclusions of the pile driving contest are:

1. Driving identical piles with four different hammers in similar soil conditions revealed considerable differences in driving characteristics.

2. Blowcount predictions reveal a wide scatter. The average was reasonably good. Variations are probably mainly due to differences in estimated driving resistance.

Most predictions were based on the same interpretation method and models. The scatter in prediction results may be related to the difference between individual users. Insufficient data were available to evaluate the importance of interpretation methods used in the prediction contest.

3. The Delmag D-25 was operating poor at low soil resistance levels while driving was fast at higher soil resistances.

4. The ICE D-640 was driving fast and well controlled at low to medium resistances. A gradual increase from low to high blowcounts occurred in the bottom sand layer.

5. The hydraulic hammers IHC SC-40 and MHF 3-3 were operating well at low energy levels resulting in medium blowcounts. The blowcounts were medium to high in the bottom sand layers.

6. The maximum compressive stresses and the impact force were high for the Delmag D-25 and low for the ICE D-640. The hydraulic hammers gave values between above two hammers with somewhat higher results for the IHC-SC-40.

7. The IHC D-640 was operating at the highest blowrate. The blowrate of the other hammers was about 40 to 60 bl/min. The blowrate of the Delmag D-25 was lower in the bottom sand layer.

8. Measurements of tensile stresses were poor.

9. There appears to be little or no correlation between the energy level for a certain hammer and the maximum compressive stress in the pile.

10. The interpreted dynamic soil resistance appears to be hammer dependent. This

resistance may be affected by the blowrate or the shape and duration of the impact.

11. The models for interpreting dynamic soil resistance and static soil resistances are insufficient. Further studies including effects of blowrates and multiple blows should be performed.

12. The driving records suggest that a higher blowrate has a positive effect on the drivability at low to medium resistances.

REFERENCES

Beringen, F.L., van Hooydonk, W.R. & Schaap, L.H.J. 1980. Dynamic Pile Testing - an Aid in Analysing Driving Behaviour, *Seminar on the Application of Stress Wave Theory on Piles*, Stockholm.

Goble, G.G. & Rausche, F. 1976. Wave equation analyses of pile driving-program manuals. Department of Transportation, Report No. FHWA IP-76-14.3.

Rausche, F., Goble, G.G. & Moses, F. 1971. A new Testing Procedure for Axial Pile Strength. *Offshore Technology Conference, Paper No. OTC 1481*, Houston, Texas.

Smith, E.A.L. 1960. Pile Driving Analysis by the Wave Equation. *Jrnl of the Soil Mech. and Found. Eng. Div., ASCE, Vol 86, No. SM4*.

APPENDIX: FORCE-TIME HISTORIES

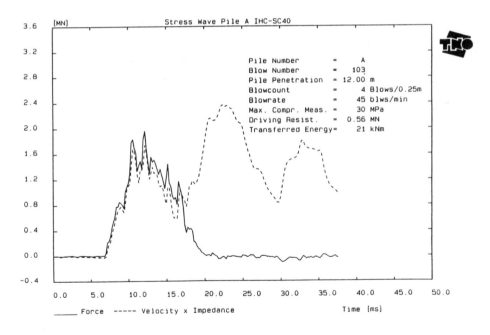

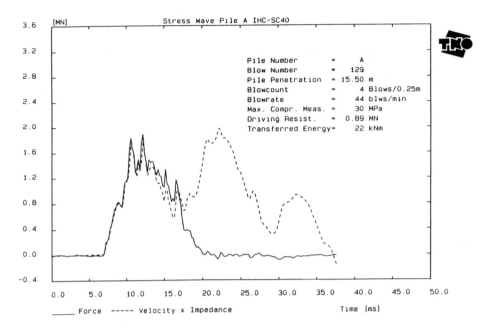

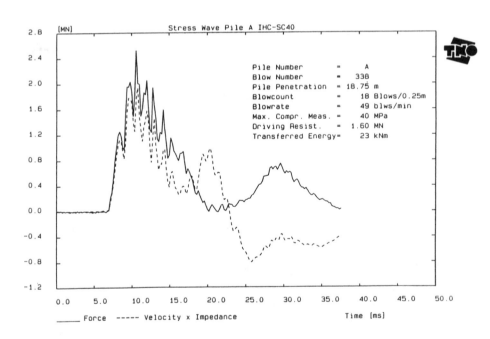

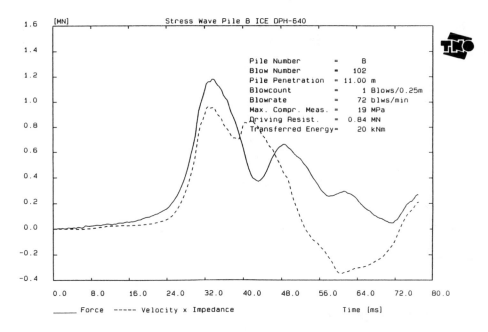

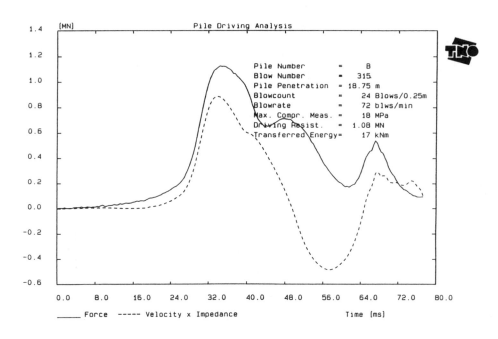

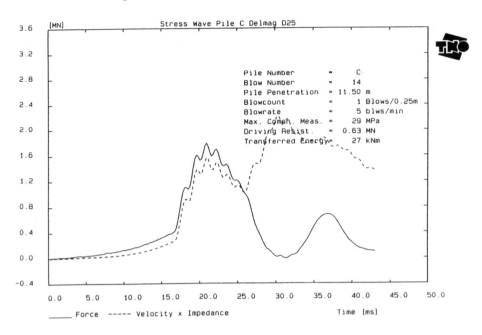

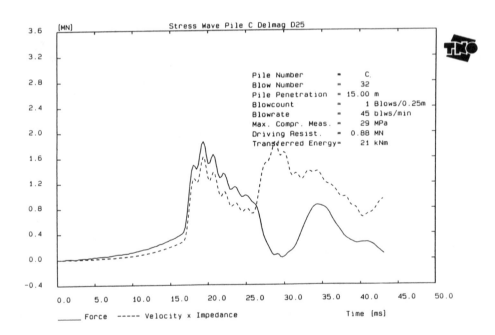

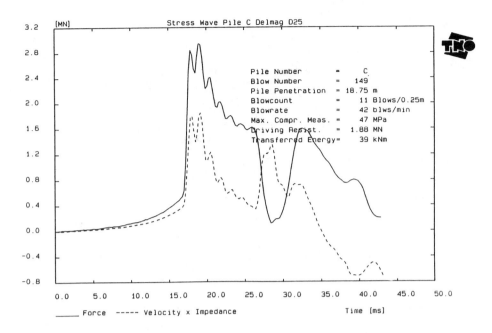

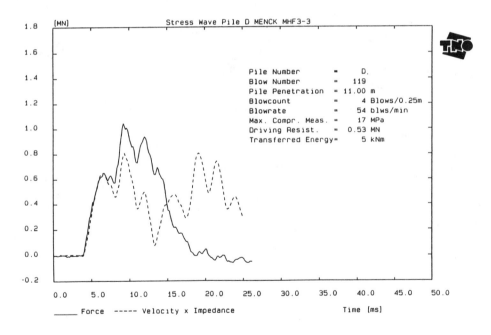

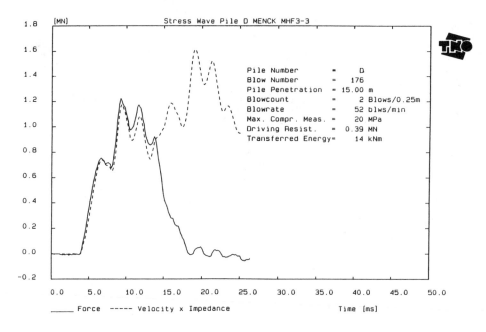

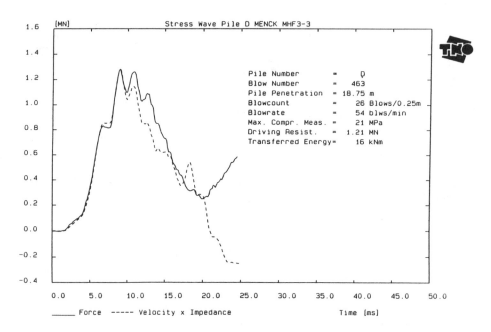

CHAPTER 6

Pile driving demonstration

D. ARENTSEN
IHC Hydrohammer BV, The Netherlands

ABSTRACT: Since the introduction of hydraulic hammers, there is still a lack of comparison information between diesel hammers and hydraulic hammers. Each of two diesel hammers and two hydraulic hammers drove at the same location an identical pile to the same penetration. The piles were monitored by Pile Dynamic Analyser (PDA) of TNO while observing the hammers. The results were used to compare the characteristics of both types of hammers. The results were also compared with the outcome when using the Hiley formula with input data, collected during driving.

6.1 INTRODUCTION

Four identical piles have been driven to target depth during the Stress Wave Conference held in Delft, The Netherlands. Purpose of the contest was to assess the difference in hammer performance of diesel hammers and hydraulic hammers including two types of drive caps.

The assessment includes environmental aspects like noise and vibrations. The assessment does not consider purchase costs, exploitation costs and reliability aspects.

To compare the difference in 'operation', the four hammers started at the same time. Winner was the crew that was the first to reach final penetration without pile damage.

6.2 SOIL DATA

A soil investigation has been performed at every single pile location (see location map, Fig. 6.1). The soil report (see Chapter 3) indicates that the soil conditions are basically uniform.

6.3 DATA OF PILING EQUIPMENT

Two diesel hammers and two hydraulic hammers were selected with a ram mass within the range of 2.5-3.0 tons (see Table 6.1).

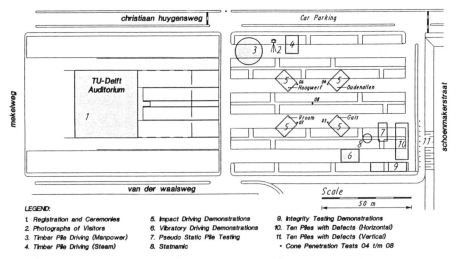

Figure 6.1. Location map.

Table 6.1. Basic data of the four piling contractors and used hammers.

Piling company	Oudenallen	Hoogwerff	Guis	Vroom
Piling rig	Kobelco 7045 XLR	Hitachi KH150-3HD	Linkbelt LS 120	Hitachi 150
Hammer type	Delmag D-25 single acting diesel	ICE DPH-600 double acting diesel	Menck MHF3-3 hydraulic	IHC SC-40 hydraulic
Max. potential hammer energy (kJ) (as published by manufacturer) Actual energy may be less	79	53	30	40
W_r, mass of ram (kg)	2500	2730	3200	2500
Drive cap type (see Fig. 6.2)	A	A	A	B

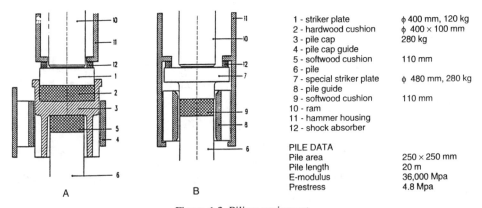

1 - striker plate ϕ 400 mm, 120 kg
2 - hardwood cushion ϕ 400 × 100 mm
3 - pile cap   280 kg
4 - pile cap guide
5 - softwood cushion   110 mm
6 - pile
7 - special striker plate ϕ 480 mm, 280 kg
8 - pile guide
9 - softwood cushion   110 mm
10 - ram
11 - hammer housing
12 - shock absorber

PILE DATA
Pile area   250 × 250 mm
Pile length   20 m
E-modulus   36,000 Mpa
Prestress   4.8 Mpa

Figure 6.2. Piling equipment.

Assumed is that the characteristics of hydraulic hammers are almost identical.

One hydraulic hammer was equipped with a different type of helmet or drive cap, without a hammer cushion . It only has a special striker plate (see Fig. 6.2 type B).

The other three hammers used a more conventional drive cap, type A, with striker plate, hardwood hammer cushion, pile cap and softwood pile cushion.

6.4 INSTRUMENTATION

6.4.1 *Hammer instrumentation*

In order to measure the velocity of the ram, both diesel hammers have been equipped with radar devices for the measurement of ram velocities, while both hydraulic hammers are standard equipped with proximity switches which measure the velocity of the ram just before the impact.

The measuring principles of the hydraulic hammers are shown in Figure 6.3.

The IHC hammer measured the time (t) between activation of sensor A and B. These sensors are located on a calibrated distance of 35 mm from each other. The ram velocity is calculated by

$$v = 3.5 \times 10^{-2} / t$$

and displayed on the control box.

The Menck hammer measured the time (t) between activation and de-activation of the single sensor. Velocity for this hammer is calculated by

$$v = 6 \times 10^{-2} / t$$

Both hammers display automatically the hammer energy, calculated according to the following formula:

$$E_{net} = \tfrac{1}{2} W_r v^2$$

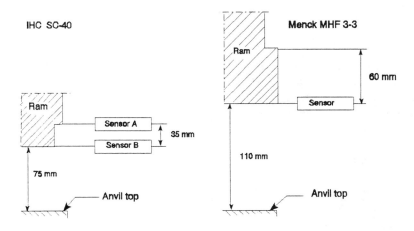

Figure 6.3. Measuring principle.

6.4.2 *Pile instrumentation*

Approx. 0.5 m below the pile head, strain gauges and accelerometers have been mounted to record the pile head stresses and the energy into the pile. Also approx. 10 m from the pile head, strain gauges have been mounted to measure the pile stresses. This signal was interrupted, due to damage of the strain gauges while penetrating the soil.

6.4.3 *Noise measurements*

Separate measurements have been made for every hammer. During these recordings, the other hammers were not in action.

6.4.4 *Recording signals*

All measurements and signals have been recorded and collected by TNO. These measurements will be kept for some years and will be available as public domain.

Part of the signals are detailed and interpreted for comparison purposes in this report.

6.5 DRIVING TRAJECT

6.5.1 *Competition traject (0.0-18.5 m)*

As mentioned before, the objective of the contest was to drive the pile as fast as possible to target depth of 18.5 m. The piling crews were free to choose the 'pre-set' of the hammer and the manner in which the piling was performed. It will be clear that these differences in execution did not contribute to a solid base for comparison.

The piling velocity was more governed by courage and swiftness of the piling crew than providing a solid data base. Therefore it could happen that one pile head started to get damaged and the crew did not want to stop for a correction. The comparison would have benefitted if corrected for such events.

6.5.2 *Measuring traject (18.5-19.0 m)*

The last 50 cm of penetration were driven one pile after the other, in order to prevent interference during noise and vibration recordings. During this traject all signals from pile and hammer sensors were collected to compare the performance and characteristics of the different hammer types and caps.

6.6 DEFINITIONS

To exclude any misunderstanding in the definition of various values, the following definitions are presented.

The most significant parameter involved in pile driving, is the energy output of the hammer.

Although most manufacturers of hammers provide maximum energy ratings (maximum potential energy) for their hammers, these values are usually downgraded by geotechnical experts for various reasons. A number of reasons such as hammer condition and wear, are known to reduce the energy output of the hammer.

E_p, potential hammer energy
The potential hammer energy E_p is the energy delivered by the hammer which can be determined during driving.

– Single acting diesel hammers: The following Delmag formula is usually used to calculate E_p:

$$E_p = \frac{4415.W_r g}{r^2}$$

where: E_p = potential hammer energy per blow (kJ); W_r = mass of ram as given by manufacturer (ton); g = acceleration of gravity (9.81 m/sec^2); r = blows per minute (1/min).

– Double acting diesel hammer: The potential hammer energy is hard to determine during operations. When no special instrumentation is supplied, this is not possible. The manufacturer does not provide a correlation between blows and energy. (For the ICE DPH no potential hammer energy data are available.)

– Hydraulic hammers: Modern hydraulic hammers as used at the test have a built-in energy measuring sensor. The energy may be read from a control box.
The formula is:

$$E_p = (\tfrac{1}{2}.W_r v^2).10^3$$

where: E_p = potential hammer energy (kJ); W_r = mass of ram (ton); v = velocity of ram just before impact (m/sec).

In case no built-in sensors are present, the E_p can be estimated by: $E_p = H.W_r$ (kJ); H = stroke m; W_r = mass of ram (t).

E_{pmax} = maximum potential hammer energy
The maximum potential energy or rated energy is stated in the brochure of manufacturer.

E_n = impact energy
The impact energy is the energy of the hammer delivered to the striking plate.

– Diesel hammers: For diesel hammers the E_n can not be determined during piling operation. A part from E_n is the kinetic energy of the ram plus some explosion energy. In soft layers, the explosion energy contributes to the penetration of the pile, but in hard layers, for example at final penetration, the effect of explosion is small.

During the test, radar was installed to measure the ram velocity for calculating the kinetic energy ($E_n = \tfrac{1}{2} W_r v^2$) of the ram.

– Hydraulic hammers: Modern hydraulic hammers are equipped with sensors to measure the ram velocity (v) just before the ram hits the striker plate.

The control box calculates the impact energy according to the following formula and shows it on the display:

$$E_n = \tfrac{1}{2}W_r.v^2.10^3$$

where: E_n = impact energy delivered to anvil or striker plate (kJ); W_r = mass of ram (t); v = ram velocity just before impact (m/sec).

For these hammers, the potential energy is in fact the same as impact energy ($E_p = E_n$).

For hydraulic hammers without sensors, E_n can not be determined during driving and consequently not displayed on the control box.

E_t = *enthru or transferred energy*

The enthru or transferred energy, is the energy measured in the pile. At the Stress Wave Conference this energy has been determined by the TNO-FPDS system (strain gauges and accelerometers) based on the following formula:

$$E_t = {}_{t0}\!\int^{t1} Fv_p dt$$

where: E_t = energy in the pile (kJ); $t0$ = time of start of impact; $t1$ = time at which the reflection of impact wave reaches the sensors (sec); F = force in the pile, based on the measured strain inside the pile (MN); v_p = the calculated velocity in the pile by integration of the pile acceleration.

η_h = *hammer efficiency*

The hammer efficiency is the ratio of the net energy and the actual potential energy.

$$\eta_h = E_n / E_p$$

where: η_h = hammer efficiency; E_n = impact hammer energy; E_p = actual potential hammer energy.

– Diesel hammers: For diesel hammers, the hammer efficiency can not be determined during operation because E_n can not be determined. During the test E_n was determined.

– Hydraulic hammers: For hydraulic hammers with sensor, efficiency η_h is about 100% because $E_p = E_n$. In case no sensors are available, the efficiency is hard to be determined.

η_c = *cap efficiency*

The cap efficiency is the ratio of enthru energy and impact hammer energy.

$$\eta_c = E_t / E_n$$

where: η_c = cap efficiency; E_t = energy per blow measured in the pile; E_n = impact energy of the hammer, delivered to the striker plate.

– Diesel hammers: For these hammers η_c is difficult to determine because E_n is normally unknown.

During the test, E_n was measured by radar, so it is possible to calculate η_c.

– Hydraulic hammers: η_c can only be determined for hammers with sensors.

η_{tot} *total efficiency*

The total efficiency is the ratio between enthru or transferred energy and potential energy:

$$\eta_{tot} = E_t / E_p$$

where: η_{tot} = total efficiency of hammer and pile cap; E_t = energy per blow measured in the pile; E_p = potential hammer energy.

– Diesel hammers: For single acting hammers, η_{tot} can be determined when enthru energy and blow rate of the hammer are measured during driving.

Using the blow rate, the actual potential energy (E_p) can be calculated.

For double acting diesel hammers without any information about the actual potential energy of E_p, the η_h is hard to determine.

– Hydraulic hammers: When the hammers are provided with sensors, instead of E_p the impact hammer energy (E_n) displayed on the control box has to be used to calculate the total efficiency by $\eta_{tot} = E_t / E_n$.

For other hydraulic hammers, the estimated E_p has to be used.

6.7 RESULTS

6.7.1 *Competition traject*

For this more informal part of the day, the four hammers started at the same moment. It was a noisy happening and after about 9 minutes, discussions could start.

Contractor Hoogwerff with a double acting diesel hammer ICE DPH-600 drove the pile to 18.5 m penetration in the shortest time. Figure 6.6 presents the different driving times and total blows.

6.7.2 *Measuring traject*

During this part of driving, the piles were driven after each other. Unfortunately not all signals could be collected. For instance, the radar mounted on the Delmag diesel hammer failed.

Therefore it was not possible to measure the impact velocity of the ram of the most widely used diesel hammer.

At the same penetration level of the four piles, one representative blow was selected to show the time force diagram, pile stresses, transferred energy etc. (see Figs 6.4-6.7). Other data were collected in Table 6.3.

During the compilation and comparison of the measurements, average values are used for blow count, blow rate and impact energy, whereas the values for transferred energy, driving resistance, etc.) are taken from one representative blow. This may lead to some inaccurate data, as derived from the measurements. The compilation of the radar measurements could only be done by hand, which is time consuming. For that reason, only one blow of the ICE hammer (see Fig. 6.7) has been used for the calculation of the transferred energy.

The signal of the radar device and the derivation to E_n are not included.

During the judgement and comparison of the results, trends rather than absolute values were evaluated. The parameters 1-8 in Table 6.3 have been used to obtain the values for the derived parameters 9-14 for comparison of the four hammers and two types of drive cap.

Table 6.2. Driving times and total blows.

Piling crew/ ranking pos.	Hammer type	Time	Total blows
Hoogwerff 1	ICE DPH-600	4 min : 4 sec	323
Oudenallen 2	Delmag D-25	6 min : 22 sec	152
Vroom 3	IHC SC-40	7 min : 27 sec	344
Guis 4	Menck MHF3-3	8 min : 48 sec	474

Table 6.3.

Item	Parameter		Unit	Delmag D-25	ICE DPH-600	Menck MHF3-3	IHC SC-40
0	Total blows			154	323	476	344
1	Blow no.			149	315	465	333
2	Blow rate	(r)	bl/min	40	72	52	44
3	Blow count		bl/0,25m	11	24	26	18
4	Impact energy per blow	(E_n)	kJ	–	20.5	23	30
5	Transferred energy	(E_t)	kJ	25	13	16	18
6	Max. compression stress		MPa	31	14	21	33
7	Driving resistance		MN	1.27	0.89	1.11	1.42
8	Noise emission		dB(A)	93.4	95.9	96.7	88.1
9	Potential energy	(E_p)	kJ	67	–	23	30
10	Set per blow	(s)	mm	22.7	10.4	9.6	13.9
11	Set per kJ	(E_t)	mm/kJ	0.91	0.80	0.60	0.77
12	Hammer efficiency	(E_n/E_p)	%	–	–	100	100
13	Cap efficiency	(E_t/E_n)	%	–	–	69	60
14	Total efficiency	(E_t/E_p)	%	37	–	69	60

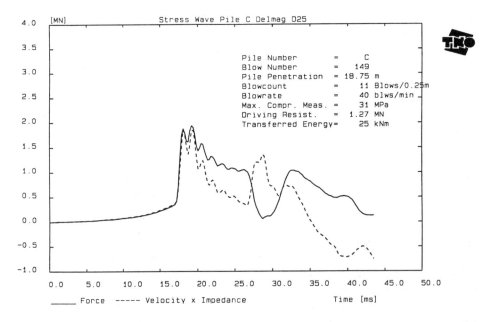

Figure 6.4. Stress wave Pile C Delmag D25.

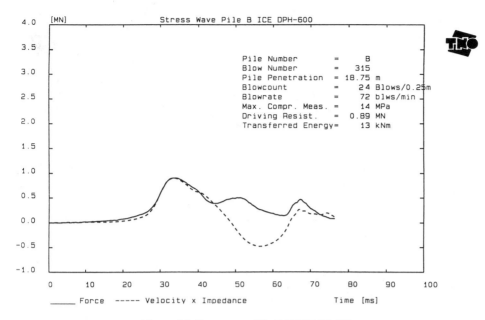

Figure 6.5. Stress wave Pile B ICE DPH-600.

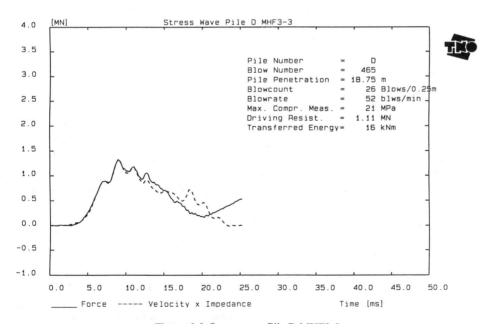

Figure 6.6. Stress wave Pile D MHF3-3.

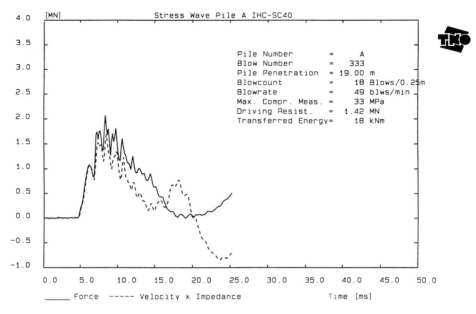

Figure 6.7. Stress wave Pile A IHC-SC40.

6.8 PILING FORMULAS

Although it is well known that piling formulas or its variations are not always correct for the calculation of driving resistance or bearing capacity, they are still often used or specified for the acceptance criteria of the driven pile. The test is also an opportunity to compare the driving result based on a piling formula.

Mostly used, is the Hiley piling formula:

$$R_u = \frac{F.W_r H}{s + 1/2\,(C_1 + C_2 + C_3)} \cdot \frac{W_r + e^2 W_p}{W_r + W_p}$$

where: R_u = ultimate bearing capacity of pile; s = set per blow; C_1 = elastic compression of cushion and cap; C_2 = elastic compression of pile; C_3 = elastic compression of soil; W_p = pile mass; W_r = ram mass; H = drop height of ram; e = coefficient of restitution; f = efficiency.

To eliminate the discussion about efficiency and other not easy determined factors, the term

$$f \cdot \frac{W_r + e^2 W_{p-}}{W_r + W_p}$$

which is more or less an overall efficiency, can be replaced by a hammer factor K.

Further $W_r H$ can be replaced by the actual hammer energy (E), which can be observed during driving. That is E_p for diesel hammers or E_n for hydraulic hammers.

The elastic deformation $C_1 + C_2 + C_3$ can be replaced by one factor C, which can be determined by paper and pencil on the pile.

Because the bearing capacity R is calculated from observed figures during driving, it is more realistic to replace R by the static driving resistance R_u. At a redrive, few weeks after driving, R_u is close to R.

Taking the above into consideration, following simplified version of the Hiley formula can be used:

$$R_u = \frac{K.E}{s + 1/2C}$$

where: R_u = Static resistance during driving (MN); K = Hammer factor, including all efficiencies and effects. This can be derived from the results of the pile monitoring by PDA and hammer monitoring. For diesel hammers $K = K_d$, for hydraulic hammers $K = K_h$; E = Hammer energy in kJ. For diesel hammers this is E_p as derived from the blow count. For hydraulic hammers this is the measured impact energy E_n; s = Penetration per blow (mm); C = Total elastic deformation of pile and soil.

Because diesel hammers have for long been present in the market, the value for the acceptance criteria of driven piles with the use of a driving formula is more familiar.

Even an efficiency of 100% for a diesel hammer is often used for the determination of K_d.

The value of K_h for hydraulic hammers often represents a problem, due to the lack of references in certain areas.

In these cases the ratio K_h/K_d at the determination of K_h can be useful.

All the test piles have been driven in the same soil to the same target depth during the test. Consequently R_u is equal to all four piles. Because the values for s, E_p for D-25 and E_n for MHF and SC-40 are known, the ratio K_h/K_d can be calculated in combination with an assumed value for 'C' (not measured).

In Table 6.4 these calculated ratios for MHF3-3, SC-40 and D-25 are given for the various values for C (mm).

Within the inaccuracy of the driving formula, here a ratio for concrete piles is about:

$$K_h/K_d = 1.5$$

The hammer factor for the various hammers can also be calculated from the measured values of R_u, S and E and an assumed value for C (mm). The results are given in Table 6.5.

With regard to
– the inaccuracy but simplicity of the piling formula,
– and taking into account the local regulations,
– the demand for simple approval criteria,
a further investigation may be suggested to the value of the hammer factor as function of pile dimensions, hammer type, and type of pile cap.

Taken into account that the by PDA measured R_u for both hydraulic hammers is not the same, it may be expected that each hydraulic hammer type has its own hammer factor, which is indicated in Table 6.5.

Table 6.4. Calculated ratios for *C* values.

Hammer combinations	K_h / K_d $C = 5$	$C = 10$	$C = 20$
MHF3-3/D-25	1.40	1.50	1.70
SC-40/D-25	1.45	1.60	1.63

Table 6.5. Calculated hammer factor.

Hammer type	E (kJ)	s (mm)	R_u (MN)	Hammer factor K C=5	C=10	C=20
D-25	E_p=67	22,7	1,27	0,48	0,53	0,62
MHF3-3	E_n=23	9,6	1,11	0,58	0,70	0,95
SC-40	E_n=30	13,9	1,42	0,78	0,89	1,13

To improve the accuracy of acceptance criteria by piling formulas, following procedure on a site is recommended:

a) Driving a 'calibrating' pile at a location of soil investigation;

b) At several penetration levels and at final penetration R_u to be measured by PDA;

c) At the same penetration level, *s* and *C* to be measured by paper on pile and pencil;

d) At the same penetration level, the hammer to be monitored for E_n or E_p, depending on hammer type;

e) Using the collected data the *K* value can be calculated and can be used for other piles to calculate R_u from measured *s*, *C* and hammer energy E_n or E_p, assuming the soil conditions, hammer, cap and piles are the same.

For relation between static driving resistance R_u and ultimate bearing capacity *R* of a pile, a static load test or semi-static load test or a redrive few weeks after driving is recommended.

6.9 CONCLUSIONS AND RECOMMENDATIONS

a) There is a considerable difference between the total number of blows per pile. Yet the hammer with the lowest total number of blows was not the first to reach target depth.

Especially in the weak upper layers, the penetration per blow by the diesel hammer was greater. This can be explained by the greater contribution of the explosion energy to the penetration in that area.

b) The representative blow is taken between six to eleven blows before reaching target depth, so virtually at the same level for reliable comparison under basically the same soil conditions.

c) The difference in driving resistance can be the result of the higher ram velocity just before impact of the hammers with a lighter ram weight (D-25 en SC-40), giving a larger damping.

d) The penetration per kJ of transferred energy is less for hydraulic hammers as compared to diesel hammers. There is no clarification yet. This may be the subject for a further investigation.

e) The higher blow rate of the ICE hammer and the fact that it reached the target depth first, (despite the high total number of blows), is in line with the experience obtained with hydraulic hammers: higher blow rate in combination with lower impact energy is often more effective than a lower blow rate at maximum impact energy (of course at equal supplied hydraulic energy).

Hydraulic hammers with independently controllable blow rate and impact energy, have proven in practice to provide the smoothest hammer operation and optimal pile penetration.

The relationship between blow rate and pile penetration per blow at constant power absorption has not formed part of any calculation so far. This may be the subject for a further investigation.

f) The total efficiency related to potential energy, shows the difference between hydraulic and diesel hammers which is in line with practice, where the total efficiency varies between 30 and 35% for diesel hammers and between 60 and 70% for hydraulic hammers on concrete piles.

g) Because the measurement of the impact velocity at the D-25 failed, its impact energy E_n remained unknown and hence also the cap efficiency. From experience and other measurements, one can expect a hammer efficiency between 40 and 50% for a diesel hammer in good condition.

This means that the impact energy E_n during the test must have been between 27 and 33 kJ, with a cap efficiency of between 92 and 76%.

The same cap will probably give a better cap efficiency under a diesel hammer (D-25) than under a hydraulic hammer (MHF3-3), which is also in line with the practice.

Possibilities to improve the cap efficiency for hydraulic hammers may be the subject for a further investigation. In any case it is incorrect to assume that a well operating cap for diesel hammers will also perform well with a hydraulic hammer.

h) The difference in cap efficiency of cap type A and B may be explained by the difference in impact velocity of the ram weight. From E_n and the mass of the ram weight, the impact velocity can be derived as being:

$v = 3.8$ m/sec for MHF3-3
$v = 4.9$ m/sec for SC-40

It has not been proven that a smaller mass of cap parts (400 kg at MHF3-3 versus 280 kg at SC-40) gives a better cap efficiency.

Also the lack of a hammer cushion at the SC-40 has no clear effect on the cap efficiency.

i) The difference in noise emission between diesel hammers and hydraulic hammers with a separate cap (type A) is small. Measurements indicate that an enclosed cap (integrated in the hammer (type B), results in less noise.

One can assume that a hydraulic hammer (with closed energy system), in combination with an enclosed cap, is more suitable for noise reducing measures than a diesel hammer.

This assumption is confirmed by the appearance in the market of hydraulic hammers with provisions to reduce the noise emission during concrete pile driving to noise limits accepted in urban areas.

j) The stress wave diagrams (Figs 6.4-6.7) show that each hammer has its own 'signature'.

Please note that the time scale for. 6.05 differs from the other diagrams.

The difference in signature is difficult to explain on a single measurement only. The difference between MHF3-3 and SC-40 can be caused by the different pile cap and the lower impact velocity on the smaller cap mass of the SC-40. The difference between D-25 and DPH-600 is partly explained by the large difference in transferred energy per blow: 25 kJ and 13 kJ resp. Also the difference in measured maximum compression in the pile can be explained by the differences in transferred energy per blow and impact velocity.

k) When using a piling formula like:

$$R_u = \frac{K.E}{s + 1/2C}$$

the hammer factor K can be calculated from the results of PDA, hammer monitoring and elastic deformation of pile and soil.

l) For concrete piles the hammer factor K_h for hydraulic hammers is about 1,5 times hammer factor K_d of diesel hammers, when impact energy E_n of hydraulic hammers and the potential energy E_p of diesel hammers is taken into account.

m) It may be an exception that the hammer factor K_h is not the same for all hydraulic hammers. More investigations are recommended.

CHAPTER 7

Vibratory pile driving techniques

G. JONKER
International Construction Equipment BV

7.1 INTRODUCTION

The object of the vibratory pile driving demonstration was not only to show the capabilities of this type of hammer to drive sheet piles, pipe piles etc. but also to show the differences between high frequency hammers and the more conventional medium frequency hammers.

It was also the intention to show the differences in soil- and pile behaviour during the driving of sheet piles as well as their extraction, not only in terms of soil resistances but also in terms of ground vibration levels.

Furthermore the advantages were shown of a newly developed hammer which did not have the resonance features during the start and stop phases of conventional vibratory hammers.

Another object of the demonstration was to show different possibilities of clutch monitoring systems.

7.2 PRINCIPLE OF OPERATION

As vibratory piling hammers function fundamentally different than impact hammers it is thought to be of importance to explain in this section the principle of operation of vibratory pile driving hammers and the differences in soil behaviour in comparison with conventional impact piling hammers.

The most common configuration of a vibratory hammer is given in Figure 7.1.

The oscillating motions (vibrations) are generated in the gear-case of the vibratory hammer.

The gear-case contains an even number of rotating shafts to which eccentric masses and gears are connected. The gears synchronise the rotation of the eccentric masses. Due to this synchronisation in each pair of eccentrics one of them rotates clockwise and the other counter clockwise.

This results in a cancellation of the horizontal components of the centrifugal force and a summation of the vertical ones (see also Fig. 7.2).

The generated vertical forces result in a vertical sinosoidal motion which amplitude (generally given as the peak to peak value) is a function of the total eccentric moment and the total mass in motion.

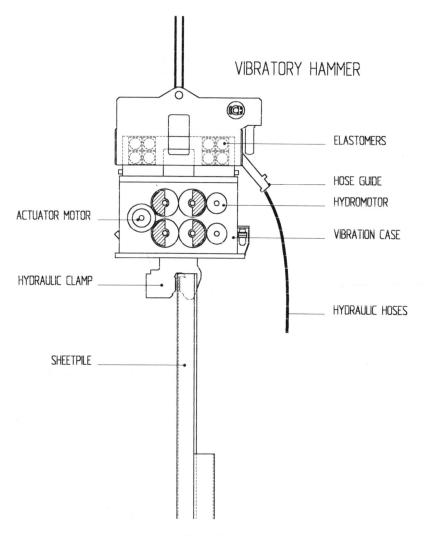

VIBRATORY HAMMER

ELASTOMERS

HOSE GUIDE

HYDROMOTOR

ACTUATOR MOTOR

VIBRATION CASE

HYDRAULIC CLAMP

HYDRAULIC HOSES

SHEETPILE

Figure 7.1

The oscillating gear-case is on top connected to the suppressor housing by means of a set of rubber, (elastomer) springs. The suppressor and springs avoid the transfer of energy (and vibrations) to the crane(boom) and functions also as a static bias weight to increase both the achievable penetration and the penetration rate.

The clamp or clamps holding the (sheet)pile are rigidly connected to the foot-plate of the gear-case. When all bolts are tightened correctly and proper clamp-jaws are used, there is no loss of energy when the forces are transferred from gear-case to pile.

As vibratory hammers are a less wellknown piece of equipment than impact hammers, a listing is made of the main differences between an impact hammer and a vibratory hammer.

– An impact hammer is a high-peak force, low-frequency hammer, whereas a vibratory hammer combines a low force with a high frequency;

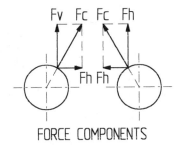

FORCE COMPONENTS

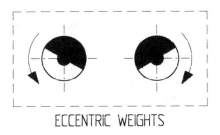

ECCENTRIC WEIGHTS

Figure 7.2.

– A vibratory hammer has a rigid connection between hammer and pile whereas with an impact hammer there is a no-tension connection;

– A vibratory hammer supplies its energy continuously to the soil, whereas an impact does this intermittantly.

7.3 VIBRATORY HAMMERS USED DURING THE CONFERENCE

The vibratory hammers used on and prior to the demonstration day of the conference are all hydraulically powered and provided with a single universal clamp, suitable to grip over the clutch of double sheet piles.

During the last 15 years important developments have taken place in the design and performance of vibratory hammers.

First there was the development of a range of 'high frequency' hammers, operating at frequencies ranging from 30 to 60 Hz.

The boundery line between medium frequency (MF) and high frequency (HF) is not exactly defined but is generally taken as 30 Hz.

To minimize the nuisance of (ground) vibrations the vibratory hammer manufacturers developed hammers with a 'resonance free' start and stop as well as a 'variable eccentric moment' feature (see next section). Also a tandem vibratory hammer was developed where the two hammers were synchronised in such a way that the vibration wave generated by one hammer was compensated by the other. This was the so called DI-POLE vibratory hammer.

The make and main characteristics of the used vibratory hammers are listed in Table 7.1. A more detailed listing is presented in Table 7.2.

Table 7.1

Hammer	Type	Max.moment (kgm)	Max.frequency (RPM)
ICE-416	MF	23	1600
WW-PV 75	HF	9	3400
ICE-26RF	RF	26	3200
ICE DI-POLE	MF	$2 \times 11,5$	1600

Table 7.2. Technical data vibratory hammers.

	Unit	ICE-416	ICE-26RF	WW-PV 75	ICE DI-POLE
Eccentric Moment	kgm	23	0-26	9	2 x 11,5
Max. Frequency	Hz	26,6	50,0	53,3	26,6
Max. Centrifugal Force	kN	650	0-1250	1010	2 x 325
Dynamic Weight (incl. clamp)	kg	3450	3850	2700	2 x 1500
Total weight (incl. clamp)	kg	6000	9780	6800	4300
Max. Amplitude (p-p)	mm	13	11	7	16
Max. Oil flow	l/min.	320	750	450	2 x 176
Max. oper. pressure	bar	340	340	320	310
Max. Power	kW	181	330	240	202
Inst. Power (Powerpack)	kW	220	369	300	220

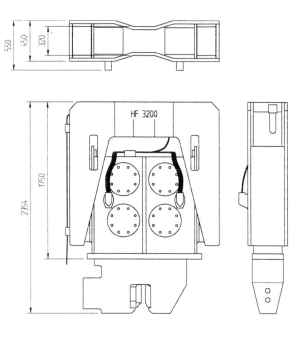

HAMMER WW - PV 75

Figure 7.3.

ICE-26RF

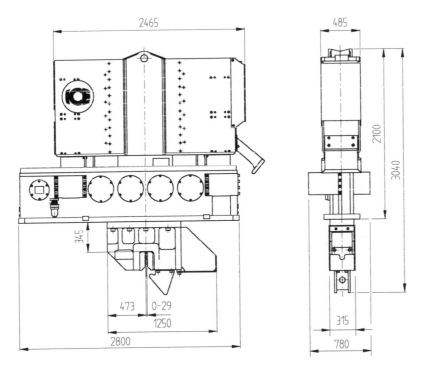

Figure 7.4.

The main dimensions and outlining of the WW-PV75 and ICE-26RF are given on Figures 7.3 and 7.4.

The ICE-416 is a conventional hammer combining a large amplitude with a medium frequency.

The WW-PV 75 is a modern high frequency hammer combining a relatively low eccentric moment with a low weight and a high maximum frequency.

The ICE-26RF (RF stands for Resonance Free) is the first operational hydraulic vibratory hammer with variable eccentric moment whereby two eccentric weights are positioned on one shaft. The eccentrics could be shifted relative to each other by remote control to enable an eccentric moment setting from 0 to 26 kgm at varying frequencies from 30-50 Hz, resulting in a maximum centrifugal force of 1250 kN.

7.4 MINIMIZATION OF GROUND VIBRATIONS

The main limitations for the use of vibratory hammers has always been the generated ground vibrations which can be a nuisance for persons and which could cause cracks in walls or floors in adjacent buildings. The vibrations could also lead to uncon-

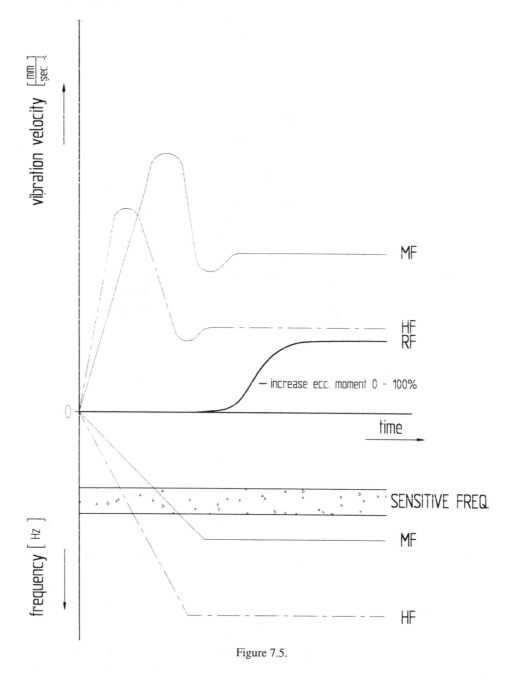

Figure 7.5.

trolled settlements of soil in the immediate surrounding of the piling operations.

The peak of the generated vibrations always occurs during the very short periods in which vibratory hammers are started and stopped.

This by reason of the fact that the operating frequency of vibratory hammers (25-60 Hz) lies above the 'sensitive' or 'resonance' frequency of most soils (15-20 Hz) and this sensitive frequency-band has to be passed when the hammer is started and when it is stopped.

The peak-vibration level when passing these frequencies can be up to 600% higher than the vibration levels measured during operation at the working frequency of the hammer.

This phenomenon is illustrated in Figure 7.5.

To eliminate the afore mentioned peak-vibrations, the hammer manufacturers have equipped their hammers with a device whereby the rotating eccentric weights can be shifted relative to each other in such a way that the total eccentric moment can be varied during operations from 0 to 100%. In the '0'position, the individual eccentric weights are positioned in such a way that not only the horizontal components of the centrifugal forces are cancelled but also the vertical ones (see Figs 7.2 and 7.6).

When starting a hammer the eccentric moment is first set in the '0'-position. When the eccentric weights have reached their normal operating frequency the moment is increased until its maximum eccentric moment or until a maximum prescribed (ground) vibration level is reached.

The differences of the two systems in terms of (maximum) ground vibration levels is illustrated in Figure 7.5.

The ICE-26RF is built such that each shaft carries 2 eccentric weights, which can be shifted relative to each other. This principle has the advantage that at the same time that the eccentric moment is decreased the operating frequency can be increased to maintain the design maximum centrifugal force and to be able to use at each setting the maximum installed power in the powerpack. This is illustrated in Figure 7.7.

Another principle to minimize the ground vibrations is to create a system whereby two identical vibration waves, but in counter phase propagation, are generated. The vibrations generated by the first hammer will be cancelled by those of the second hammer.

In practice this means that one has to design a vibratory hammer system built-up out of 2 identical hammers, being positioned as close as possible to each other to ensure minimum vibrations.

This resulted in the ICE DI-POLE; a hammer made of 2 ICE-216 medium frequency hammers being synchronised by a cardan shaft to ensure a phase difference of 180 degrees at all times. A lay-out of the ICE DI-POLE is given in Figure 7.8.

To prove the principle of operation a number of tests were carried out at the ICE premises whereby the hammers were not connected with the cardan shaft to make it possible that one hammer was running slightly faster than the other.

This results in a situation where the hammers are sometimes completely in phase and sometimes completely out of phase.

This could be clearly measured by geophones. One of the results being presented as Figure 7.9.

As a conclusion it can be stated that the principle of the DI-POLE works excellent however the disadvantage of the system being that only 2 single sheet piles can be

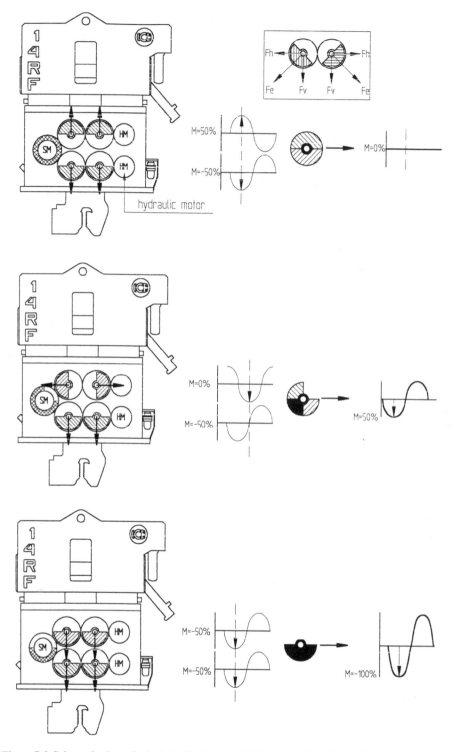

Figure 7.6. Schematic view of principle of virbrating driving, using four identical excenter weights.

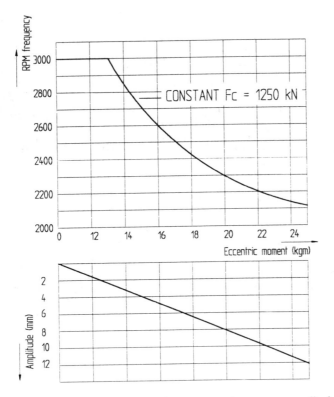

Figure 7.7. Dependency between (static) excenter moment and frequency or amplitude for a resonance free vibrator.

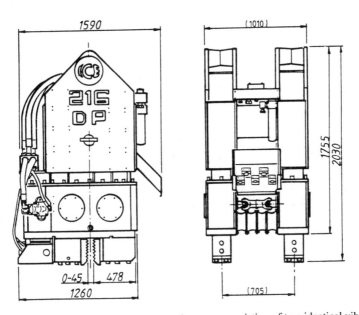

Figure 7.8. View of the D1-Pole Vibratory hammer, consisting of two identical vibrators.

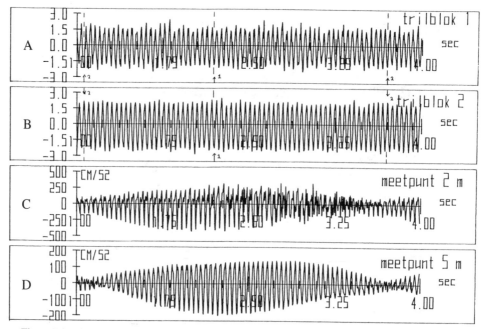

Figure 7.9. A and B: Amplitude vs. time for both vibrators.
C and D: Accelerations vs. time at two different observation points at ground surface.

driven and that a new configuration of the hammer is required to convert them for two sheet piles of another size or another make.

Although it is proven that the principle works at all frequencies, the single 'resonance-free' type hammers have shown to be more convenient in practice.

7.5 SHEET PILE INSTALLATION

Appr. 3 months prior to the demonstration day, a number of ARBED (ISPC) sheet piles types PU20 and AZ26 had been installed. The length of the piles was 20 m and the installation was carried out by general contractor Woud Wormer using an ICE-416.

The piles were installed prior to the conference to have the possibility to get a comparison of the driving and extraction behaviour during driving and extracting of piles with and without being subjected to soil set-up. The piles were installed in one row as indicated in Figure 7.10.

The main technical data of the sheet piles are as listed in Table 7.3.

Some typical time count diagrammes (time required per 1 m penetration) are given in Figure 7.11.

The term 'time count' has been introduced to have an analogue to the blow count of impact driven piles. A high time count indicates that more time is required to penetrate a certain distance identical to a high blow count indicating that more blows are required to drive the pile .25 m or 1 ft.

It can be seen that even for such long sheet piles driving was relatively easy and

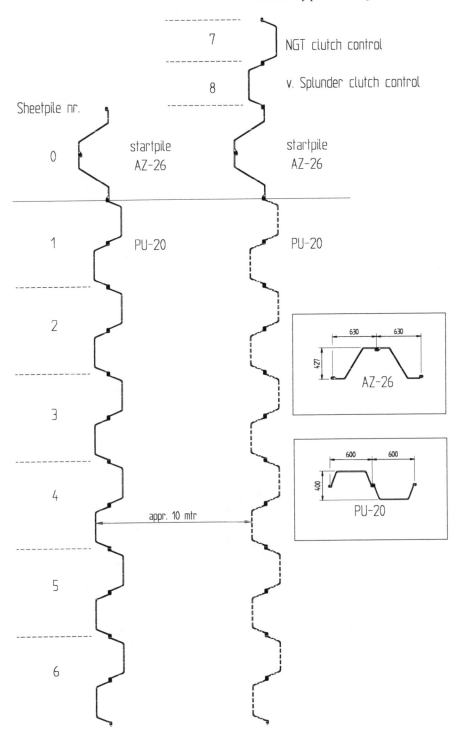

Figure 7.10.

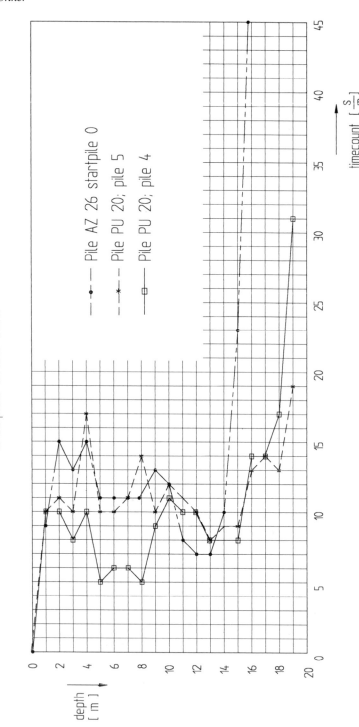

Figure 7.11.

Table 7.3. Technical data sheet piles.

	Unit	ARBED PU20	ARBED AZ26
Profile width (double)	mm	1200	1260
Profile depth (double)	mm	400	427
Moment of Res. (single)	cm^3	480	1640
Moment of Res. (double)	cm^3	2400	3280
Moment of Inertia (single)	cm^4	7080	34970
Moment of Inertia (double)	cm^4	48050	69940
Cross section (single)	cm^2	108	125
Cross section (double)	cm^2	216	249
Weight (double)	kg/m	169	196

Table 7.4

Sheet pile no.	Hammer manuf.	Ecc. mom. Rel.	Ecc. mom. Abs. (kgm)	Frequency rel.	Frequency Abs. (H2)	Centr. force (kN)	Remarks
1	WW-0V75	low	9	high	53,3	1010 kN	
2	ICE-26RF	low	10	high	50	1010 kN	
3	ICE-26RF	high	26	low	31,6	1010 kN	
4	ICE-26RF	low	18	low	31,6	1010 kN	
5	ICE-26RF	var.	0-26	var.	30-60Hz	1250 kN	
6	ICE-Dipole		2.11, 5		26,6	2 × 235kN	
7	WW-PV75		9		53,3	1010 kN	NGT clutch control
8	WW-PV75		9		53,3	1010kN	Splunderclutch control

was never more than 1 minute per meter and in total took never more than 4 minutes.

7.6 PROGRAMME DURING THE DEMONSTRATION DAY

The original anticipated programme during the demonstration day was to extract the sheet piles one by one (as doubles) and install them again appr. 10 m away from their first position as indicated by the dotted sheet piles in Figure 7.10.

Two sheet piles were equipped with different methods of pile-interlocking detection switches, namely sheet piles numbers 7 and 8.

To enable a study to the influence of different eccentric moments and frequencies the following scheme as given in Table 7.4 was anticipated.

Due to several reasons the programme as listed above could not be executed during the demonstration day, the main reasons being:
– Slight delay in the demonstration day activities prior to the vibratory hammer demonstrations;
– More set-up than expected causing long extraction times;
– Problems and consequently further delay with one of the clutch-monitoring systems;
– Change-over from one vibratory hammer to the other took longer than expected.

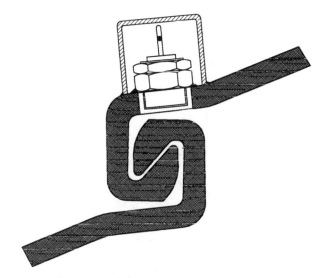

Clutch detection system van Splunder

Figure 7.12.

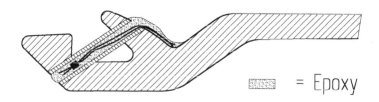

= Epoxy

Clutch detection system NGT

Figure 7.13.

It was on-site decided to leave most of the piles in the ground and demonstrate only the clutch-interlock-monitoring systems, as well as the main feature of the ICE-26RF vibratory hammer with its resonance free start and stop mechanism.

7.7 CLUTCH DETECTOR MONITORING SYSTEM

The two systems demonstrated in Delft had the following principles:

The VAN SPLUNDER-system uses a proximity switch near the lower end of the sheet pile to be driven. This proximity switch gives a positive signal if the two sheet piles still interlock with one another and a negative signal if this is not the case anymore. This principle is shown on Figure 7.12.

In the NGT-system one or a number of break pins are installed in the clutch of the sheet piles already driven. The break pins are part of a closed loop electric circuit.

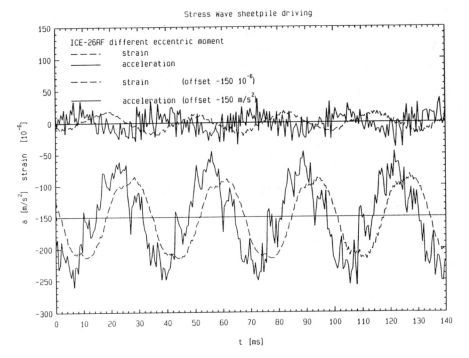

Figure 7.14.

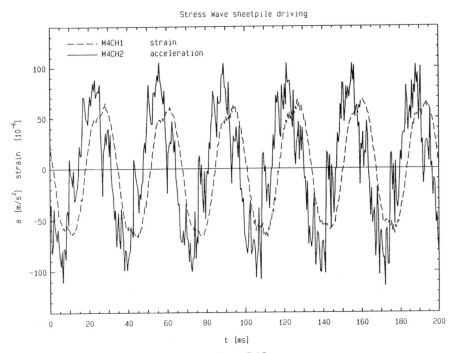

Figure 7.15.

When the break pin is cut through by the next sheet pile the circuit is broken and this is monitored at ground level (see also Fig. 7.13).

Contrary to the VAN SPLUNDER-system the NGT-system gives only information on the levels where the pins are installed, whereas the VAN SPLUNDER-system gives a continuous signal over the total installation length.

As proven on many sites both systems work well and are an additional safety on critical sites.

7.8 DYNAMIC MEASUREMENTS

Due to the limited available computer storage capacity as well as the very limited time available at the end of the demonstration day, only a few measurements could be made on the sheet piles.

They are presented as Figures 7.14 and 7.15. Type of sheet piles extracted was Arbed PU20 (as doubles).

The results shown indicate the same operating frequency (appr. 33 Hz) however at different eccentric moment settings.

It also shows that the acceleration signal is less sensitive to stress-waves of the second and third order than the strain signal.

Advertisements

The testing programme was made possible by the "Stichting Onderzoek Kwaliteit Paalfunderingen" (Foundation for Research on Quality of Pile Foundations) and Dutch Government financial support. The Foundation's objectives are to instigate, execute and supervise research on the quality of pile foundations, including their environmental effects. The research includes foundation design, pile instrumentation and associated interpretation methods, and installation techniques.

The Foundation is formed by the following six Dutch companies:

Ballast Nedam Engineering B.V

Fugro Engineers B.V.

ICE B.V.

IFCO Funderingsexpertise B.V.

IHC Hydrohammer B.V.

Nederhorst Grondtechniek B.V.

FUGRO: YOUR WORLDWIDE ASSOCIATE FOR INSTRUMENTATION SERVICES

- Geotechnical
- Structural
- Environmental
- Onshore
- Nearshore
- Offshore

DESIGN PHASE
- Special soil sampling and in-situ testing
- Laboratory testing
- Pile load testing
- In-situ foundation model testing
- Environmental monitoring

CONSTRUCTION PHASE
- Settlement monitoring
- Sonic integrity testing
- Pile and hammer monitoring
- Noise and vibration assessments

IN-USE PHASE
- Foundation behaviour monitoring
- ROV inspection
- Structural monitoring
- Machinery monitoring

Fugro Engineers B.V. is a member of the Fugro group with offices throughout the world which provide geotechnical, environmental and surveying services.

Fugro Engineers B.V.
10, Veurse Achterweg
P.O. Box 250, 2260 AG Leidschendam
The Netherlands
Tel.: 31-70-3111444, Fax.: 31-70-3203640

Advanced Pile Testing Equipment

for quality assurance in piling

TNO has created two new Foundation Pile Diagnostic Systems; the FPDS-5 and the FPDS-α, in addition to the existing FPDS-4 system. The main advantage of all TNO's Foundation Pile Diagnostic Systems is their Multi-functional design. Multiple testing options can be run on a signal system, making it very cost effective.

FPDS-5 is a data acquisition system consisting of a standard notebook computer of the customers choice and a high-quality signal conditioning system. The FPDS-5 complements the existing line of FPDS systems. FPDS-4 is better adapted for users who work in harsh or aggressive environments and who still need a powerful DOS computer. Both FPDS-4 and FPDS-5 offer the following available options:

FPDS-4

Integrity testing	Pile Driving Analysis
Dynamic Load Testing	STATNAMIC™
Vibration Monitoring	Radar Hammer Monitoring
Vibratory Driving Monitoring	SPT Controller
Hydraulic Hammer Monitoring	Seismic Cone Monitoring
Sway Monitoring	Thickness measurements

FPDS-α is a microcomputer based upon a Motorola 68332 microprocessor. It is a lower cost unit than in FPDS-4 and FPDS-5, but remains of high quality for what it is designed to do. FPDS-α options are:

Pile Driving Controller	Vibration Monitoring
Vibratory Driving Controller	Inclination Monitoring

TNO has a full range of state-of-the-art stress wave program prediction and signal matching software, called TNOWAVE. TNOWAVE offer the following applications:
- Sonic integrity testing, automatic signal matching
- Dynamic load testing, automatic signal matching
- Impact hammer pile driving prediction
- Vibratory hammer pile driving prediction
- Statnamic simulation

FPDS-5

TNO since 1932
Representatives Worldwide

TNO Building and Construction Research
P.O. Box 49
2600 AA Delft
The Netherlands
Telephone: +31 15 284 22 72
Fax: +31 15 284 39 96

TNO has more than 20 years experience in pile testing and manufacturing pile testing equipment. We have been at the leading edge in the development of all new and innovative pile testing techniques. We are backed by the resources of TNO, one of the world's largest independent research organizations with a staff of over 4000.

making technology work

IFCO's main area of consultancy is foundation engineering for buildings that are still in the planning stage. For this service it is of utmost importance to be able to predict the future behaviour of the foundation. It is equally as important to be able to verify the predicted foundation behaviour by measurements. To achieve a leading role in this area, IFCO has invested in the design and use of specific measuring equipment. The experience that has been gained with this strategic concept, has proven correct. IFCO now has a lot of experience combined with a large amount of data from the measurements, giving its consultancy a solid base. So far, the following measurement equipment has been developed by IFCO:

IFCO Foundation Expertise BV

P.O. Box 334
2800 AH Gouda
The Netherlands
·Telephone xx31 - 182 536 000
Faximile   xx31 - 182 571 242

IFCO IT-system

The IFCO Integrity Testing-system (IT-system) is used to detect irregularities and/or cracks in concrete pile shafts and determines the exact location of the fault. The system is operated by one person only and can normally test more than 60 piles within an hour.

IFCO VM-system

The IFCO Vibration Monitoring-system (VM-system) measures vibrations automatically and continuously in three directions. The system works on standard batteries (for a period of one month) and warns automatically when a preset level is exceeded.

IFCO PDA-system

The IFCO Pile Driving Analysis system (PDA-system) is used to measure the stress and the acceleration of the pileshaft during each blow of the piling process. With this measuring system it is possible to check the bearing capacity of piles. The data from the measurements also gives an important contribution to the prevention or solution of possible piling problems. The same equipment can be used for carrying out a Dynamic Load Test.

IFCO PD-system

The IFCO Pile Driving Documentation-system (PD-system) allows the rig operator to monitor each pile without delay and provides him real-time with detailed information about the dynamic soil resistance, hammer efficiency, etc. The system requires one small accelerometer and it is possible either to fix the computer permanently in the piling rig or to use it as a mobile monitoring system. The IFCO PD-system can be easily extended to a full PDA-system.

Other IFCO-measurement systems are:
the IFCO Optical Displacement-system
the IFCO Load Displacement-system
the IFCO Data Acquisition-system
For more information, please contact us.

a solid base

PILING EQUIPMENT NEDERLAND B.V.

FUNDEX GROUP

A dynamic group specialised in the many facets of foundation technique, piling, drilling and manufacturing of piling equipment.

The experience the company has gained from working in foundation construction has been translated into the development of a range of piling equipment, designed to provide the customer with the most suitable piling rig for his work.

For further information please contact:
Fundex Group,
Brugsevaart 6,
P.O. Box 55,
4500 AB OOSTBURG,
The Netherlands

Telephone : **31 117 457510
Telefax   : **31 117 457560

As a piling contractor the Fundex Group has experience in:

- Vibrex: cast in situ piles
- Fundex: vibration-free soil displacement piles
- Tubex: vibration-free steel casing piles which can be used in areas of limited working height
- Bored piles: to large diameters
- Diaphragm Walls: using a Deep Trench Cutter (DTC)
- Slurry Walls
- Deep shafts
- Steel Casings
- Anchor Piles
- Pseudo Static Pile Load Testing (PSPLT)

PSPLT

GEOTECNICA CIENTEC S.A.C

Consultores de Ingeniería

SAFE AND COST - EFFECTIVE SOLUTIONS

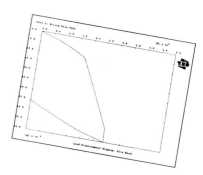

● **DYNAMIC TESTS ON PILES & FOUNDATIONS**
  - Sonic Integrity Tests **(SIT)***
  - Dynamic Load Tests **(DLT)***
  - Pile Driving Analysis **(PDA)***
  - **STAT*NAMIC®* *** Load Tests
  - Vibration Monitoring **(VIBRA)***

● **GEOTECHNICS**
  - Soil Mechanics for Foundations, Dredges & Landfills
  - Rock Mechanics for Foundations & Mining
  - Borings, Site Investigations, SPT & CPT Tests
  - Land & Water Contaminations
  - Laboratory Tests

● **CONCRETE & ASPHALT CONTROL QUALITY**
● **NON-DESTRUCTIVE TESTS AND CONSTRUCTION PATHOLOGY**
● **GEOPHYSICS, TOPOGRAPHY, BATIMETRY & METHEOROLOGY**

* With License, Support and Technical Assistance from TNO - Building & Construction research (The Netherlands)

SINCE 1960 LEADING THE EXCELLENCE IN QUALITY ASSURANCE IN THE CONSTRUCTION INDUSTRY

Lavalleja 847 - Buenos aires (1414)
ARGENTINA

TEL: 54 - 1 - 862-0547/861-8126
FAX: 54 - 1 - 861-8126

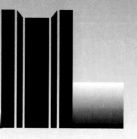

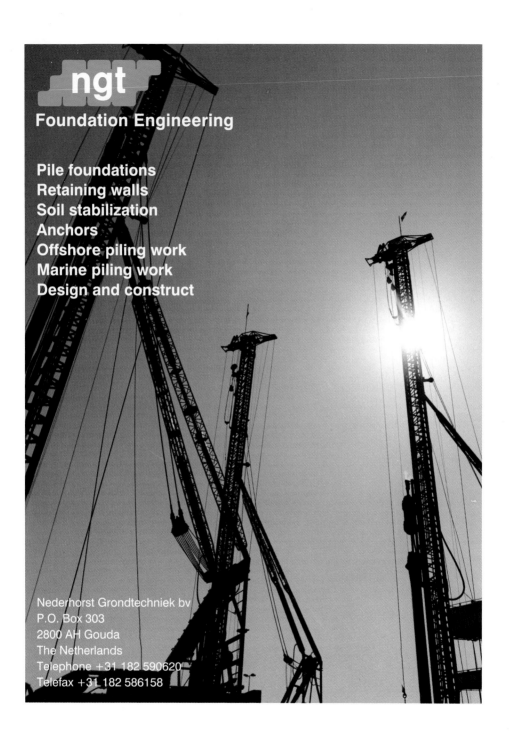

For Experienced Drivers Only

Water. A challenging environment in which to work. In which to build major structures such as jetties and quay walls, bridges and tunnels, dams and causeways, breakwaters and sea defences. Water. Not for the faint-hearted, the inexperienced. It calls for special skills, ingenuity and in-depth knowledge. Ballast Nedam has been finding innovative solutions to marine construction problems for more than a century. For clients world wide. It has the know-how and resources to take complex projects from concept through to completion. In the fields of building, civil and marine engineering and dredging. On land. In water. The challenge continues.

Laan van Kronenburg 2
P.O.Box 500
1180 BE   AMSTELVEEN
The Netherlands

Telephone + 31 20 5 45 91 11
Telefax   + 31 20 6 47 30 00

 **Hydrohammer**

Multifunctional hydraulic piling hammer:

- Steel and concrete piles.
- Useable as rock breaker and soil compactor.
- Under water driving without energy losses.
- Noise reduction to L_{Aeq} = 76 dB(A) at 15 m.
- Printed driving record.
- Performance display or control unit.
- Controllable between 5 to 100%.
- Hammers from 30 kJ to 2300 kJ.

IHC Hydrohammer on concrete piles
with sound enclosure.
Noise level at 15 m L_{Aeq} = 76 dB(A)

Driving large diameter steel piles,
up to 2500 mm diameter.
Larger diameters are possible.

IHC Hydrohammer B.V.
PO Box 26
2960 AA KINDERDIJK
The Netherlands
Phone : +31.78.69.10.302
Fax : +31.78.69.10.304